UNIVERSITY REVIEWS IN BOTANY

Editor: Professor V. H. HEYWOOD, PH.D., D.SC.

4

BIOLOGY OF
THE RHODOPHYTA

UNIVERSITY REVIEWS IN BOTANY
Editor: Professor V. H. Heywood, Ph.D., D.Sc.

Biology of
the Rhodophyta

PETER S. DIXON
University of California at Irvine

OLIVER & BOYD
EDINBURGH

OLIVER & BOYD
Croythorn House, 23 Ravelston Terrace,
Edinburgh EH4 3TJ
A division of Longman Group Limited

ISBN   0 05 002485 X

Printed in Great Britain by
T. & A. Constable Ltd, Edinburgh

CONTENTS

PREFACE

The Rhodophyta is a very specialised group of plants, with thalli often of beautiful symmetry and with somewhat elaborate reproductive structures and life histories. Despite their many attractive features, the group is often omitted completely or given only the most cursory treatment in textbooks so that students complain that the red algae are far too complicated and difficult to understand. The present text is an attempt to overcome these complaints. In a treatment as brief as this, it will be inevitable that some readers will find items omitted which they consider should have been included or that the balance between topics is not what they might have wished to find. The major problem is preparing this text has been to decide what could not possibly be left out rather than to determine what had to be included. In addition, there are many aspects of critical importance for which the available data are either unsatisfactory or lacking completely. The present text represents a brief survey of the Rhodophyta, of the problems if not the answers, of the areas where information is available or where it is wanting, of those aspects of the red algae which I find most interesting. It is offered in the hope that through it others might find something of the fascination and attraction which the red algae have for me.

Literature published prior to December 1971 has been considered in some detail although only information of the most critical significance published after January 1972 has been incorporated into this text.

I am indebted to Dr Isabella Abbott and Dr Mark Littler for reading the manuscript and for their advice, comments and helpful criticism. I also wish to thank Dr L. R. Blinks, Dr A. D. Boney, Prof. M. De Valera, Dr G. F. Elliot, Mrs L. M. Irvine, Dr M. Martin, Dr S. N. Murray, Dr M. Parke, F.R.S., Dr G. F. Papenfuss, Dr W. N. Richardson, Dr J. L. Scott and Dr J. West for help, assistance and information during the preparation of this publication.

INTRODUCTION

The red algae are a very specialised group of plants differing markedly in many respects from the representatives of other algal divisions. One of the major characteristics of the Rhodophyta is the presence of phycobilin pigments, of which the red phycoerythrin usually predominates. Despite their name, the blend of pigments which they contain can produce a whole range of colour variation, from blue-green through red, black or purple. In addition, photodestruction of the phycobilin components gives a more greenish or yellowish aspect to the thallus while certain red algae, regarded either as partially or completely parasitic, contain reduced quantities of photosynthetic pigments or none at all. Irrespective of the actual colour of the plant, the same predominant storage polysaccharide ('floridean starch') occurs throughout the group. A second major characteristic feature is the complete absence of all motile stages, both in male gametes as well as in all the many forms of spore which occur. There is no evidence that motile stages have ever occurred since true centrioles appear to be lacking from the nuclei. Early reports of flagellated male gametes were due to the misidentification of the motile stages of fungi contained within the algal cells, while the supposed unicellular flagellate representatives such as *Rhodomonas* have been transferred to other groups.

Except for a few very specialised cases, the cell structure is similar throughout all the red algae. The cell wall is composed of randomly arranged microfibrils embedded in an amorphous matrix of a mucilaginous nature, while the cell walls of many representatives are calcified. All members of the Corallinaceae in the field have walls which are impregnated in this way although this might not be true of some specimens grown in laboratory culture. Calcification occurs regularly in genera of families other than Corallinaceae, such as *Liagora* and *Galaxaura*, while occasional incrustation has been reported in various other genera. Despite their similar structure, cell shape and size vary considerably not only between species but also in the same species depending upon age and position in the thallus. In young newly formed cells the cytoplasm is dense and viscous. During the process of cell enlargement it becomes highly vacuolate, usually with a single large central vacuole surrounded by the cytoplasm which forms a lining to the cell wall. Thus, in mature cells, the various organelles and metabolic products contained in the cytoplasm are located in a peripheral position in the cell. The cells are all eucaryotic, with the number of nuclei present ranging from one, in most cases, to several thousand at the other extreme. A few specialised cells of different structure found in certain genera are thought to be associated with such phenomena as iridescence and halide metabolism while the function of the special hair-like cells is completely unknown. Attachment structures may be formed from single cells which form simple rhizoids or from larger cell masses which in certain cases can function like the tendrils of higher plants although neither occurs in some red algae.

Despite the many features common to all red algae, the group as a whole can be divided into two major subdivisions which differ considerably from each other in terms of their morphology, reproduction and life histories. In terms of morphology, the Florideophyceae possess pseudoparenchymatous thalli made up by the aggregation of filaments which are formed almost exclusively by the division of apical cells. Intercalary cell divisions of a specialised type may occur, particularly in certain genera of the Delesseriaceae and Corallinaceae. Most Florideophycean thalli are based on a heterotrichous organisation although this does not occur in the genera of the Ceramiales and in certain very reduced and supposedly 'parasitic' genera in other orders. Thalli in the genera of the Bangiophyceae are much more simply organised. Unicellular and colonial examples occur in this class, together with simple or branched filaments. The most advanced types are those with true parenchyma, as in *Porphyra*, or a rudimentary type of heterotrichous organisation as in *Erythrotrichia*. Cell division in thalli of the Bangiophyceae appears to be diffuse in most cases although restriction of divisions to a particular cell can occur in some cases.

In terms of reproduction, the differences between the two classes are even more marked than with their morphology. The Florideophyceae are exclusively oogamous. The zygote which is formed in each carpogonium following fusion with a single non-motile male gamete is retained on the female gametangial thallus for some time. It undergoes a series of developments, often of some complexity, to give rise to a multicellular structure referred to as the carposporophyte. This may or may not possess an independent photosynthetic apparatus but in all cases the carposporophyte would appear to be dependent to some extent on the female gametangial phase on which it has arisen.

Ultimately the carposporophyte gives rise to one or more spores and, because of its origin from a single-celled zygote and its termination in the production of one or more spores, it is frequently thought of as a morphological phase equivalent to the gametangial phase. Sporangia of many different types occur in various genera of the Florideophyceae, with mitotic nuclear divisions in some cases and meiotic divisions in others. Although the Bangiophyceae are often said to possess a sexual reproductive process similar to the oogamy of the Florideophyceae, the evidence for this is far from satisfactory. As with the latter group, spores of various sorts are formed in members of the Bangiophyceae although there is much confusion as to the origin and nuclear characteristics of these.

Life histories in the Rhodophyta, as in all divisions of the algae, have had a great fascination for investigators over many years. Despite this, little attention has been paid to the evidence required to establish a life history or to the critical evaluation of the information available. Much of the 'evidence' generally accepted is of very doubtful status and it has become accepted largely because of uncritical repetition. Critical examination of the data will permit the acceptance of several types of life history in the Florideophyceae, even though there are some aspects for which the evidence is not as good as one might wish. For the Bangiophyceae on the other hand, with the uncertainties regarding gamete fusion and the nature of the various spores which have been reported, there is not one entity for which the life history can be outlined with any degree of certainty.

The economic uses of red algae cover various activities, from products of major industrial importance to relatively minor sources of food in underdeveloped countries. The two major uses of red algae are either as human food or as the raw material from which complex polysaccharides such as agar can be extracted. There are a few reports of the use of red algae as animal fodder, although these are relatively minor compared with the extensive use of brown algae for this purpose. The only use of red algae as soil dressing is in connection with the use of coralline algae as a source of lime.

As with most algal groups, the fossil record for the Rhodophyta is poor, totally inadequate to provide any insight into their evolution. The fossil coralline algae are probably the best known of the fossil Rhodophyta because of their involvement in the formation of coral reefs, both during the recent and more ancient periods in the history of the earth.

Despite the poor fossil record, phylogenetic considerations have been numerous, both about the internal relationships as well as the relationships with other algal divisions. These phylogenetic studies have been based principally on the morphological aspects of the recent forms and are highly speculative, although the results of this speculation have been used on occasions in systematic treatments of the Rhodophyta. The systematics of the red algae have been dominated for many years by the views of Kylin and his students although, as will be shown, there are serious objections to many parts of the currently accepted system.

Table 1. *Distribution of marine and freshwater genera and species of Rhodophyta based on the data of Kylin (1956) and Bourrelly (1970)*

	Marine		Freshwater	
	Genera	Species	Genera	Species
BANGIOPHYCEAE				
Porphyridiales	2	24	5	11
Goniotrichales (see *Note 1*)	5	9	5	6
Bangiales	7	60	2	2
Compsogonales	0	0	2	10
Rhodochaetales	1	1	0	0
FLORIDEOPHYCEAE				
Nemaliales	35	399	8	122
Gelidiales (see *Note 2*)	11	70	0	0
Cryptonemiales	106	900	1	1
Gigartinales	81	732	1	1
Rhodymeniales	34	185	0	0
Ceramiales	249	1341	6	11
Ceramiaceae	70	381	3	4
Dasyaceae	14	105	0	0
Delesseriaceae	65	255	1	3
Rhodomelaceae	100	600	2	4
UNCERTAIN AFFINITY	15	17	0	0

Note 1: included in Porphyridiales in the present treatment.
Note 2: included in Nemaliales in the present treatment.

Most Rhodophyta are marine, occurring principally in the rocky intertidal or sublittoral of open coasts, although a few occur in freshwater situations. Marine representatives occur throughout the world, from the arctic and antarctic waters to the tropics. The freshwater Rhodophyta appear to be most frequent in temperate climates although they are not entirely lacking either from tropical locations or from high latitudes and there are several which occur only in damp terrestrial habitats. Certain marine species are able to extend some distance into estuaries or other brackish waters. None appears capable of withstanding frequent rapid changes from full seawater to freshwater although relatively slow acclimatisation between the two has been demonstrated for *Bangia*. Although the greatest number of freshwater Rhodophyta occurs in the Nemaliales, all orders except Gelidiales (*sensu* Kylin), Rhodymeniales and the monotypic Rhodochaetales have at least one freshwater representative. Conversely, every order except Compsopogonales has marine representatives. The very approximate distribution of marine and freshwater genera and species is shown in Table 1, based on data from Kylin (1956) and Bourrelly (1970). The system adopted in the present text differs slightly from the treatments of those authors in that two orders accepted by them are merged into other orders. The Goniotrichales is merged with the Porphyridiales and the Gelidiales with the Nemaliales.

In this brief treatment of the red algae, it is proposed to discuss the topics of structure and morphology, reproduction and life histories, with minor treatments of the fossil representatives, economic utilisation, and systematics. Those physiological topics for which there is no evidence to indicate differences from other plants will be omitted although it should be appreciated that this rejection is more often based on ignorance of processes in the red algae than on positive information as to similarity.

1
Cell Structure and Function

Except for a few minor details, the cell structure is of remarkable uniformity throughout the red algae. The cell wall is composed of randomly arranged microfibrils embedded in an amorphous matrix, both components being composed largely of polysaccharide. In a few cases, the wall is impregnated with calcium carbonate. The cytoplasm is dense and viscous in young cells, although vacuolation and enlargement result in the mature cell possessing only a thin peripheral layer of cytoplasm in which the various organelles, metabolic products and the nucleus or nuclei are located. The presence of phycobilin pigments is one of the major characteristics of the red algae which, despite their name, may exhibit a colour from among a whole range of variations due to the different blends of pigments which they each contain. Energy is absorbed principally by the phycoerythrin rather than by the chlorophyll, but it is transferred through phycocyanin to the chlorophyll where it then participates in the normal photosynthetic process. The same predominant storage polysaccharide, floridean starch, occurs throughout all the red algae. Cell specialisation occurs only in a few genera where it is associated with iridescence, or halide metabolism. Attachment structures may be formed either from single cells, which form simple rhizoids, or from larger cell masses which can function in a few cases like the tendrils of higher plants. Neither are formed in some cases and the thallus appears to attach by 'cementation'.

The cell wall

Chemical composition

The principal organic constituents of the cell wall in Rhodophyta are carbohydrates, although minute quantities of proteins or lipid materials may also be present. The carbohydrate constituents may be divided

1

approximately into two parts, an inner layer largely composed of material which is not, or only slightly, extractable by boiling water and an outer portion which is more readily soluble. The spatial differences in chemical and physical properties are more obvious in some species than in others. The presence of cellulose in red algal cell walls was first claimed by Kolkowitz by means of microchemical tests and subsequent studies confirmed this claim. By means of suitable extraction procedures, Naylor and Russell-Wells (1934) were able to identify cellulose by macrochemical methods and also to estimate the content on a dry weight basis. The figures obtained ranged from 2·1% (*Rhodymenia palmata*) to 15% (*Corallina officinalis*). Kylin (1915), on the other hand, claimed that no cellulose was present in the Bangiophyceae and this has been substantiated. As a consequence of recent developments in carbohydrate chemistry, the use of X-ray diffraction procedures by which carbohydrates in a crystalline state may be recognised and the use of electron microscopy to demonstrate the occurrence of fibrils and their arrangement, knowledge of the physical construction and chemical composition of cell walls has advanced rapidly in recent years. The X-ray diagrams of the cell walls of many red algae (Cronshaw *et al.*, 1958; Myers and Preston, 1959) are similar to, but not absolutely identical with those for cellulose II, the so-called 'regenerated' cellulose, which is the crystalline modification formed when cellulose is precipitated from solution. As Kreger (1962) indicated, this similarity cannot be regarded as positive identification because in no case could it be claimed that the X-ray figures were exactly identical. Furthermore, when the microfibrils are hydrolysed, a mixture of various monosaccharides is formed and although glucose is the major component other sugars may be present in quantity. The basic unit of the microfibrils in *Porphyra* is mannose with some xylose (Cronshaw *et al.*, 1958) so that chains are a mannan rather than cellulose, confirming the earlier statement of Kylin.

The water-soluble components in red algal cell walls, which are often described very appropriately as 'mucilages', may form as much as 70% of the dry weight of the wall. Because of their gel-forming properties, some of the red algal mucilages have been used extensively for centuries (Boney, 1965). The best known of these are agar, extracted from species of *Gelidium* and other genera, and carrageenin, extracted from species of *Chondrus* and *Gigartina* (see p. 219). There is often confusion between the crude extract and the polymer of defined constitution. It has been suggested that the suffix -*in* be used to denote the crude extract and the suffix -*an* the specific polymer (O'Colla, 1962). As a consequence of their commercial value and of the variation in the properties of the products obtained from different species, analytical studies of red algal mucilages are more advanced than for any other group of algae. It was appreciated by 1953 that galactose was the principal component in all the red algal mucilages examined and that this occurred in the form of a galactan sulphate. It is now known that both agar and carrageenin are mixtures of several complex polysaccharides (see p. 220). Although the elucidation of the detailed structures of these is progressing, some are still far from certain. In general, the mucilaginous polysaccharides of red algae are of three major types: those based on agarose-type linkages, those based on K-carrageenin linkages, and xylans. In addition a whole host of other mucilages have been extracted, from a wide range of species, whose composition has not been investigated in any detail (O'Colla, 1962).

Structure

In general it can be stated that the cell walls of all the Rhodophyta which have been examined show the same basic structure of randomly arranged microfibrils embedded in an amorphous matrix of a mucilaginous nature. The detailed structure of the microfibrils may be very similar to cellulose in most Florideophyceae or mannan in those Bangiophyceae which have been studied. The microfibrils are usually more dense in the area adjacent to the cell cavity. At the surface of the thallus there is often a 'cuticle' apparent, particularly in electron micrographs. This is usually thought to be pectic in nature and it would appear to be made up of several contrasting layers of material. The cell wall in many Rhodophyta is clearly stratified, both in the apical cell and in the derivative cells formed by its division. The detailed structure of the wall has attracted some attention, although not as much as might be anticipated in view of the phenomenal expansion which can occur (see p. 103). It has been claimed by Chadefaud (1962) and Priou (1962), on the basis of histochemical staining and optical microscopy, that in filamentous Florideophyceae the wall is made up of several major components. The filament is said to possess an outer membrane while each cell has a cylindrical wall ('locula'), adjacent cells being separated by H-shaped interlocular pieces. In thalli of Ceramiaceae, the interlocular pieces are said to be short, merely overlapping the upper and lower ends of the loculae whereas in members of the Acrochaetiaceae the interlocular pieces are described as being so elongate that adjacent H-shaped pieces meet to envelop completely each locula. Although the photographs presented appear to validate the conclusions, supporting evidence for this organisation does not appear to be forthcoming from electron microscope studies.

The outermost boundary of the thallus is an area about which little is known. With optical microscopy it is possible to demonstrate frequently that a thin 'cuticle', about 1 μm in thickness, is present. Using electron microscopy it is possible to show that the 'cuticle' is made up of several layers, usually three in number. With old tissue, the 'cuticle' is often of a yellowish colour and it is obviously composed of a fine textured material of considerable strength. The sculptured casts made by epiphytic diatoms may be observed long after the diatom cell has gone. Early views as to the chemical composition of the 'cuticle' were summarised (Fritsch, 1945) as 'probably a denser layer of pectic material'. More recent French work has identified this material as 'chitin' although the reasons for such an identification are far from clear. The presence of mannan has been demonstrated in both *Porphyra* and *Bangia* although the most recent investigation has shown that in *Porphyra* the 'cuticle' is composed of about 80% proteinaceous material (Hanic and Craigie, 1969). It is now clear that the chemically isolated 'cuticle' fractions of green, brown and red algae are all highly proteinaceous. It is interesting to note that in red algae 4-hydroxyproline is absent from the detached cuticle. Hydroxyproline is an amino acid which may characterise higher plant cell wall proteins although not of the walls of either brown or red algae (Gotelli and Cleland, 1968). Because the cuticle of higher plants does not contain protein (Roelofsen, 1959) the use of the term in the red algae is obviously open to question. There *is* an external bounding layer in the red algae but further studies are required before the term cuticle is accepted without question or a new term devised.

Calcification

The cell walls of certain Rhodophyta are impregnated with $CaCO_3$. All members of the Corallinaceae growing in the field have thalli which are calcified in this way and the same phenomenon occurs regularly in other genera such as *Liagora*, *Galaxaura* and *Peyssonelia*. It is interesting to note that with various coralline algae growing in culture the degree of calcification is frequently much reduced and in certain cases it may even be impossible to detect any calcium carbonate. In addition, there are a few reports of the occasional deposition of calcareous material in genera which are normally not calcified, as in specimens of *Nemalion helminthoides* containing a heavy endophytic infestation of *Calothrix*, or in certain of the freshwater Rhodophyta such as *Batrachospermum*.

Anhydrous calcium carbonate exists in two crystalline forms, calcite and aragonite. The former is rhomboidal and the latter orthorhombic. The two forms of calcium carbonate differ markedly in specific gravity, hardness and solubility. Many animals deposit a mixture of the two types, but this never seems to occur with the algae. In the Rhodophyta, all members of the Corallinaceae deposit calcite, whereas in *Peyssonelia* and those genera of the Nemaliales which are calcified, the calcification is always in the form of aragonite. On occasions, investigators have claimed that the other crystalline form occurred but, without exception, reinvestigation has failed to substantiate the claim. The occasional calcification mentioned previously is of unknown composition, but in the *Nemalion* infected with *Calothrix* it was probably in the form of hydrated calcium carbonate ($CaCO_3.6H_2O$). The deposition of calcareous material appears to occur first in the outer, mucilage-rich, regions of the wall and then to spread into the wall towards the area where microfibrils are more dense.

In addition to the calcium carbonate, both magnesium carbonate and strontium carbonate may occur. In algae, the occurrence of the two accessory carbonates bears some relationship to the type of calcium carbonate present. The proportion of magnesium carbonate in calcitic species is much higher than in those where aragonite is deposited. Thus, the skeletal material in *Liagora* (aragonite) contains about 1% magnesium carbonate, whereas in the Corallinaceae (calcite) there may be from 7% to as much as 30% present. The situation with strontium carbonate is exactly opposite to the relationship with magnesium carbonate; strontium carbonate is incorporated to a greater extent into aragonite than into calcite. In the Rhodophyta, the calcitic Corallinaceae surveyed contained between 0·25 to 0·37% strontium carbonate. None of the aragonitic red algae appears to have been analysed, although the content of strontium carbonate in aragonitic green algae can be as high as 2·3%. For all calcitic organisms, it has been claimed that there is a relationship between the magnesium content of the skeletal material and the temperature of the water in which it is living, although this relationship is by no means as distinct for the calcitic red algae as it is for some other organisms. Even so, the relationship has been said to indicate seasonal temperature changes, the ratio of magnesium to calcium in the calcite of *Lithothamnium* is 2 to 4% higher in the summer deposited material than in that laid down in winter (Chave, 1954). More recent studies (Moberly, 1968) have suggested that the magnesium content is a reflection of growth rate rather than simply of temperature.   Growth rate depends on various factors in addition to temperature.

The distribution of the calcareous Rhodophyta and the chemical composition of their calcareous deposits are closely related to the amount of calcium carbonate dissolved in seawater.   The level of saturation of calcium carbonate in seawater is influenced strongly by temperature and by the partial pressure of carbon dioxide.   Thus, tropical waters, particularly at the surface, are usually saturated or even super-saturated with calcium carbonate, while the level of saturation may be less than 90% in colder, polar waters (Vinogradov, 1953). Although coralline algae occur at all latitudes, the massive growths which result in 'coral reefs' (see p. 57) are almost entirely restricted to tropical waters.   Extensive subtidal accumulations have been reported in deep water in the arctic (Kjellman, 1883).

Pit connections

It has been postulated for many years that one essential difference between the Bangiophyceae and the Florideophyceae is the occurrence in the latter of structures which were referred to as 'pit connections'. These were defined (Fritsch, 1945) as 'well-marked pits, occupied by broad cytoplasmic strands, that occur in the septa between adjacent cells'.   However, there were also several subsequent reports of the occurrence of these structures in certain genera of the Bangiophyceae so that it was suggested that pit connections might not have the systematic significance formerly attributed to them (Dixon, 1963a).   The structure of these pit connections was the subject of much controversy. The first synthesis of the conflicting opinions (Jungers, 1933) indicated that there were, in the Florideophyceae, two very different structures. These were termed the '*Polysiphonia*' type and the '*Griffithsia*' type. The former was described as two densely-staining plates separated by a membrane without open communication between the two cells, while the '*Griffithsia*' type consisted of a lens-shaped structure without a closing membrane so that direct communication between the cells was possible.

The structures in question were at the limit of resolution of optical microscopy and detailed anatomical investigation was possible only after the advent of electron microscopy. Initial investigations using this technique (Myers *et al.*, 1959; Dawes *et al.*, 1961) appeared to confirm the distinction between the two types originally postulated. Using better fixation and thin sectioning it was subsequently shown that the pits of *Lomentaria baileyana* (Bouck, 1962) consisted of biconvex plugs, with a clearly defined membrane surrounding a region composed of material of two different densities. The region adjacent to the cells was more dense than the area sandwiched in the middle of the structure. Subsequent investigations have all substantiated this interpretation and none has been able to offer any support for the idea that open communication between cells might exist. The term 'pit connection' is therefore most inappropriate, in that the structure is neither a 'pit' nor an intercellular 'connection', although the term continues to be used in its original sense. The development of the structure in *Pseudogloiophloea* (Ramus, 1969a, b) shows that two stages are involved. The transverse wall dividing the cell develops centripetally as an annular ingrowth and when this invagination is complete, a rimmed aperture remains, bounded by the plasmalemma, through which cytoplasmic communication occurs between two cells. The plug then develops within this aperture by the condensation of material on flattened vesicles which lie parallel to one another and traverse the aperture. Following the development of the plug, the plasmalemma is still continuous from cell to cell and lies outside the limiting membrane of the plug caps. The pit connections vary considerably in size and shape (Feldmann and Feldmann, 1970; Lee, 1971a) not only between species, but also in different parts of the thallus in the same species. The diameter of the plug may range from 0·2 μm in the ultimate parts of the thallus of a Delesseriacean alga, to as much as 30-40 μm in mature parts of the axial filament in *Ptilota* or *Bonnemaisonia*. The thickness of the plug is equally variable, so that the plug ranges from the very flat disk found in species of *Chondria* to an almost spherical object. Deposits of electron-dense material are laid down in certain cases on the sides of the plug caps, in some instances forming prominent dome-shaped masses projecting into the cytoplasm on either side. Chemical analysis of the plugs has suggested that they are either a polysaccharide-protein complex, as in *Griffithsia pacifica* (Ramus, 1971) or a lipoprotein (Feldmann and Feldmann, 1970). The former analytical procedures are more detailed and precise and Ramus indicates that the polysaccharide component is neither pectin nor cellulose while the

protein fraction contains some basic amino acids. The plugs are therefore very distinct in structure and chemical composition from both
cytoplasm and cell wall. There is now no fine structural evidence to
substantiate the view that open communication normally occurs
between cells through the pit connections. There are several situations
in the Florideophyceae where optical studies indicate that the pit connections may reopen in certain conditions. During the formation of
the carposporangia in chains in *Lemanea fluviatilis* (Dixon, unpublished
observations) the cells are produced by apical cell division with pit
connections formed in the usual way but these appear to vanish at
the time of enlargement and conversion of each cell of the gonimoblast to a carposporangium. The migration of nuclei through the pit
connection has been demonstrated in another species of this genus
(Mullahy, 1952) while the processes of both sporangial and vegetative
regeneration involve the reopening of pit connections and the transfer
of considerable quantities of cytoplasm. There are no studies of these
phenomena available at the present time.

In the Bangiophyceae, pit connections have been reported on occasions in certain genera, although the reports have been reasonably
consistent in certain cases, as for instance in the *Conchocelis* phases of
Bangia and *Porphyra*, although not entirely so. It was suggested
(Drew, 1954b) that in the *Conchocelis* phase 'the resemblance to pit
connections seems to be due to the transverse walls remaining incomplete'. Similarly, in *Rhodochaete parvula*, pit connections were not
mentioned in the original description, although described subsequently
while Magne (1960) commented that he could not observe these structures in fresh material but that they became prominent after acetic
acid fixation. Because of such observations, it was suggested (Dixon,
1963a) that one explanation for the reports of pit connections in members
of the Bangiophyceae was simply due to incomplete closure of the
centripetal invagination at cell division. The few ultrastructural investigations of members of the Bangiophyceae have disclosed the existence
of structures similar to the pit connections of the Florideophyceae in
the *Conchocelis* phase of *Porphyra* (Lee, 1971a) but with no evidence of
their occurrence elsewhere (Honsell, 1963; Nichols *et al.*, 1966). These
studies give no indications of the incomplete closure of the annulus
which was suggested as an explanation for reports of the existence of
pit connections in other Bangiophyceae.

Several authors have attempted to relate the structure of the pit
connection in the Florideophyceae with the septal structures found in

certain members of the Ascomycetes and Basidiomycetes. The occurrence of cytoplasmic contact in the latter is one serious objection to such a relationship, although other authors (Lee, 1971a) have suggested the similarity would be much greater if the comparison were made with the Florideophycean structure from which the plug had dropped out.

Cytoplasm and organelles

Cytoplasm and vacuolation

Cell shape, size and structure in red algae show considerable variation not only between species but also in the same species, depending upon age and position in the thallus. The cell wall, as has been shown (p. 1), is composed of randomly orientated microfibrils composed of material similar to cellulose II, embedded in an amorphous mucilaginous matrix. In young, newly formed cells, the cytoplasm is dense and highly viscous, occupying the whole lumen with the exception of a few small vacuoles. During the considerable enlargement which is shown particularly by members of the Florideophyceae the cell becomes highly vacuolate. The cytoplasm, containing a nucleus, or several nuclei (see p. 21), chloroplasts (see p. 10), various organelles and metabolic products (see p. 20) is located in a peripheral position, firmly in contact with the cell wall. Suggestions that the degree of adhesion between cytoplasm and cell wall were due to the extension of the former into the latter (Bünning, 1935) have been confirmed by recent electron microscope studies (Scott and Dixon, 1972a, b). Crystals of various materials are often to be found in the cytoplasm (Feldmann-Mazoyer, 1941). Where only a single nucleus occurs in a cell, this is usually located at the basal pole of the cell, at least when fully developed.

In some red algae, particularly members of the Ceramiaceae, the process of vacuolation does not simply produce a large central vacuole and peripheral cytoplasmic lining to the cell wall. The most common situation is for the mature cell to contain, in addition to the peripheral lining, a single central axile strand of cytoplasm traversing the large axial cells between the pair of pit connections at the apical and basal poles. In other cases, a series of cytoplasmic processes, called 'pseudopodia', form a coronate cluster at the basal end of a cell (Phillips, 1925b; Feldmann-Mazoyer, 1941).

The cytoplasm of most red algae is highly susceptible to changes in the temperature or tonicity of the medium, although some species show a remarkable capacity for survival. Specimens of *Bangia* and *Porphyra* and other species of the high intertidal are dried out completely, every day for considerable periods, apparently without harm. This facility to withstand desiccation is useful for transmission of living material in that such dried specimens are capable of resuscitation on remoistening for several days and then of liberating viable spores. The capacity to withstand hypertonic solutions is said to run parallel to cold resistance (Höfler, 1931). The harmful effects of dilution of seawater are readily seen in the rapid decrease in marine Rhodophyta in estuarine conditions as well as the rapid reduction in number of species in the entrance to the Baltic Sea. Biebl (1937, 1962) has given surveys of the effects of various ecological stresses while there are now a number of more recent general surveys of the effects of light (Hellebust, 1970), temperature (Gessner, 1970), salinity (Gessner and Schramm, 1971) and water movements (Schwenke, 1971) on red and other algae. No further discussion of these topics will be provided here.

A cytoplasmic inclusion consisting of polygonal bodies identified as virus particles, surrounding a spherical body 1·0 to 1·5 μm in diameter has been detected in cells of *Sirodotia tenuissima* (Lee, 1971b). Normally this spherical body was electron dense throughout although on occasions an electron transparent area occurred at the centre. The polygonal particles each measured 50 to 60 nm in diameter and were laid down in crystalline arrays resembling closely the viral organising areas which have been identified in higher plant tissues. The spherical bodies were present in the apical cells and appeared to be distributed to daughter cells in the course of cell division. Other species of *Sirodotia* and the closely related *Batrachospermum* were also examined but no trace of such bodies could be detected in these. The identification of the bodies as virus particles was made solely on morphological grounds and it was not possible to effect transfer from infected to uninfected material of other species by growing the two together in the same culture vessel for a period of a month.

Chloroplasts and pigments

Cells of red algae may contain one, several or no chloroplasts, the latter condition occurring most frequently in internal tissues of compact thalli. A single chloroplast, which is often stellate in shape and axile in disposition, with a prominent central pyrenoid, occurs in every cell in certain genera of the Bangiophyceae. Parietal chloroplasts occur in other genera of that class. In the simpler Nemaliales each cell contains a stellate axile chloroplast with a pyrenoid, although in such chloroplasts the stellate arms frequently become very elongate with age to give a more parietal distribution. A single parietal chloroplast occurs in young cells of many of the more advanced Florideophyceae but this becomes increasingly lobed as the cell ages eventually breaking up into a number of band-shaped fragments. Many Florideophyceae have several chloroplasts per cell in all parts of the thallus. These are usually discoid or elliptical in shape in the young cells, although their outline may change considerably with age.

The chloroplasts of the red algae are similar to those of all eucaryotic algae in that the thylakoids are enclosed within a double membrane envelope. In Rhodophyta, the thylakoids are never stacked together but lie within the matrix of the chloroplast. Frequently the outermost thylakoid forms a circle immediately inside the chloroplast envelope. Occasionally, two such encircling thylakoids are present and in some cases it has been shown that there are no traces of encirclement. Where concentric thylakoids are present, the inner thylakoids which lie approximately parallel to one another may be formed by evagination from these. The chloroplasts of red algae contain the accessory pigments phycocyanin and phycoerythin. These phycobilin pigments have been shown to be located in granules which measure 35 nm located in a very regular manner on the outer faces of the thylakoid membranes (Gantt and Conti, 1965, 1966; Gantt et al., 1968). These granules have been termed phycobilisomes, although it is obvious that they are not present in all Rhodophyta (Scott and Dixon, 1972b). One other feature of the chloroplast in red algae in that the double membrane by which it is enclosed is not associated with endoplasmic reticulum. It appears that in all classes of the algae except the red algae, the Prasinophyceae and most Chlorophyceae, the chloroplast (with its double membrane) is enclosed within a complete envelope of endoplasmic reticulum (Gibbs, 1970). Between the thylakoids, areas of DNA appear as moderately electron dense regions containing fibrils 2·5-4·0 nm in diameter. It has been established in the Rhodophyta that these fibrils are DNA by digestion with DNAse (Bisalputra and Bisalputra, 1967).

The pigments which occur in red algae are of three principal kinds, chlorophylls, carotenoids and phycobilins. In addition there are certain special pigments found only in particular organisms, the most important of which is probably floridorubin reported initially in *Rytiphlaea tinctoria* (Feldmann and Tixier, 1947). It is not proposed to discuss all aspects of the general properties of algal pigments, which have been treated fully elsewhere (Lewin, 1962), but to restrict the discussion to the specialised aspects of the pigments of Rhodophyta.

Chlorophylls. The chlorophylls are green, highly labile, fat-soluble substances whose chemical tetrapyrrolic structure consists of a central magnesium atom and certain side groups. The chlorophyll molecule is associated with protein *in vivo*, but this association is readily disrupted. Five chlorophylls have been identified in plants and of these, only one, chlorophyll *a*, can be shown to be of universal occurrence in red algae. Chlorophyll *d* has been reported for some red algae, although it appears to be highly erratic in distribution. Chlorophyll *d* is particularly abundant in *Erythrophyllum delesserioides* while *in vivo* spectroscopic investigations indicate the presence of this chlorophyll in *Rhodochorton purpureum* and *Porphyra perforata*. For some reason, chlorophylls are much more difficult to extract from algae than from other plants, the solvent used most frequently for the latter (80% acetone) being largely ineffective. Duysens (1951) suggested that chlorophyll *d* might be acting as the 'energy trap' which would explain the decline in efficiency of photosynthesis in *Porphyra laciniata* and *Porphyridium cruentum* at wavelengths in excess of 650 nm. This explanation is no longer accepted, particularly since chlorophyll *d* could not be demonstrated in *Porphyridium*. The most recent interpretation is that chlorophyll *d* is probably only a slightly altered form of chlorophyll *a*. Even in green algae there are many additional absorption peaks around 675 nm which disappear upon extraction and are attributed to polymers. Such behaviour could well explain the idiosyncrasies of chlorophyll *d*.

Carotenoids. Carotenoids are yellow, orange or red fat-soluble substances, comprised of isoprene units, usually eight in number. There are two principal classes of carotenoids, the purely hydrocarbon carotenes and their oxygenated derivatives, the xanthophylls. The total quantity of carotenoid in red algae is much less than the amount of chlorophyll. There is usually about 3 to 10 times as much total chlorophyll as total carotenoid. The principal carotene of Rhodophyta is β-carotene, although α-carotene is said to predominate in *Phycodrys rubens* (Larsen and Haug, 1956). Of the xanthophylls, lutein is present in most of the red algae examined, while taraxanthin occurs in *Rhodymenia palmata*. Fucoxanthin has been reported from several species, but not confirmed on re-examination, and it is likely

CELL STRUCTURE AND FUNCTION   13

that these claims are a consequence of heavy diatom contamination. Several of the species concerned (e.g. *Polysiphonia nigrescens, Callithamnion pikeanum*) are often so heavily epiphytised in the field that the normal red coloration is obscured by the brown diatom pigmentation.

Phycobilins. The phycoerythrins and phycocyanins are red and blue biliproteins readily soluble in water but not in fat solvents. They are known to be photosynthetically active only in algae, and there only in the divisions Rhodophyta, Cyanophyta and Cryptophyta (O'hEocha, 1962). Phycoerythrin and phycocyanin each consist of a tetrapyrrolic phycobilin which, unlike the chlorophylls, is difficult to separate from the associated protein moiety. For this reason, most studies are of the biliprotein rather than of the free phycobilin component. There are several variants of both phycoerythrin and phycocyanin. These were once thought to be related to the class or division to which the organism from which the pigment had been extracted was referred. These variants were assigned prefixes to indicate systematic origin, as R-phycoerythrin, C-phycoerythrin or B-phycoerythrin to denote phycoerythrin obtained from members of the Rhodophyta, Cyanophyta, or Bangiophyceae respectively. The differences between the variants were largely with respect to the peaks of the visible light absorption spectrum. R-phycoerythrin has three peaks at 495, 540 and 560 nm respectively while B-phycoerythrin lacks the 495 peak and shows no clear-cut separation of the 540 and 560 nm peaks, but rather a plateau. C-phycoerythrin, obtained from blue-green algae, has a single peak but this varies from 545 to 568 nm depending on the species from which it was obtained. Within red algae, there are various problems with respect to phycoerythrins, the maximum at 540 nm, for instance, being obvious in some cases and totally lacking in others, while repeated recrystallisation (Airth and Blinks, 1956) may cause the 560 nm peak to disappear. It is possible that the explanation is simply that one is dealing with a group of three pigments of which any organism may contain all, or two, or only one, the last situation being that found in Cyanophyta so that C-phycoerythrin represents a single pigment of the trio. The table of biliprotein distribution data presented by O'hEocha (1962) indicates quite clearly that red and blue-green algae are perfectly demarcated on the basis of this phycoerythrin content, but that within red algae the distribution of R-phycoerythrin and B-phycoerythrin, identified on the basis of their absorption spectra, does not correlate at all with the systematic attribution of the organisms involved.

The position with respect to the phycocyanins is similar to that for the phycoerythrins in that R-phycocyanin has two maxima, at 550 and 620 nm respectively, whereas C-phycocyanin only has the one at 620 nm. However, there do not appear to be any differences between the phycocyanins of the two major red algal classes such as were first proposed, even if not subsequently justified, for the phycoerythrins. With the phycocyanins, the supposed R-phycocyanin and C-phycocyanin, as first suggested, are not restricted to one or the other division. Furthermore, it has been claimed by some workers that R-phycocyanin occurs in *Porphyra perforata*, whereas others have stated that it is C-phycocyanin. Because of this confused situation, it has been suggested that R-phycocyanin is a mixture of C-phycocyanin and a phycoerythrin, the latter providing the 550 nm peak. While this concept of a mixture is attractive it does not explain all the information available. Allophycocyanin is widely distributed in red algae, both of Bangiophyceae and Rhodophyceae. This has a single absorption peak at about 650 nm and a molecular weight of 135 000 compared with values of 290 000 for R- and B-phycoerythrins and 270 000 for R-phycocyanin. It was once thought that allophycocyanin was simply a degradation product because it was first extracted from commercial dried 'Nori' (= *Porphyra* species). This concept would appear to be incorrect. In view of the recent identification that the pigment involved in photoperiodic phenomena in Angiosperms is very similar to allophycocyanin and that true photoperiodism can be demonstrated in certain red algae (see p. 97), interest in allophycocyanin is bound to increase.

Floridorubin. The rhodomelaceous alga *Rytiphlaea tinctoria*, when placed in freshwater or allowed to remain in a container of seawater of inadequate size, liberates a carmine-purple pigment in profusion. The specific epithet applied to the alga when it was first described at the beginning of the nineteenth century is indicative of this characteristic, but the claim (Debaux, 1873) that the pigment represents the 'purple of the ancients', also known as 'Tyrean purple', is incorrect. That the pigment released by *R. tinctoria* differed from the red pigment characteristic of most red algae was appreciated by Kützing (1843), who named it 'phykohaematin', but it was then not studied seriously for over a century. Feldmann and Tixier (1947) showed that the released pigment was not a biliprotein, that the name applied by

Kützing was a gross chemical misidentification and suggested that floridorubin was a more suitable name for it. They also claimed that phycoerythrin did not occur in *R. tinctoria*. Floridorubin was subsequently recognised in four other species of the Rhodomelaceae, *Vidalia obtusiloba*, *Rhodomela confervoides*, *Polysiphonia nigrescens* and *Lenormandia prolifera*. The last detection (Saenger *et al.*, 1969) showed that phycoerythrin *was* present in addition to floridorubin. It had been stated (Feldmann and Tixier, 1947) that not all the floridorubin in *Rytiphlaea* was extracted by the procedures adopted and that some remained firmly attached to the membranes. Using a different extraction procedure the presence of both floridorubin and phycoerythrin has been demonstrated in *Lenormandia prolifera* and also in *Rhodomela confervoides* so that it seems likely that the pigment reported as being firmly attached to the membranes was phycoerythrin. The purified floridorubin has an absorption spectrum with maxima at 285 nm and 525 nm. The pigment changes reversably to yellow if the pH is lowered and the aqueous extract is not decomposed by boiling. In these respects, floridorubin is very different from phycoerythrin which is changed irreversibly by both of these treatments. The pigments of those members of the Rhodomelaceae previously reported as containing only floridorubin need to be reinvestigated. The occurrence of floridorubin has been suggested as being a criterion of systematic importance; this aspect is discussed elsewhere (p. 246).

The physiological role of floridorubin is obscure. As phycoerythrin is present also, at least in certain species, it is not likely that floridorubin has any significance in the photosynthetic process. It has been shown (Saenger *et al.*, 1969) to undergo reversible oxidation and reduction with changing pH but there is no evidence that it functions as a redox compound in the cell.

Photosynthesis

Photosynthesis is a complex series of reactions by which the fixation of carbon dioxide is effected through the use of light energy, although the process involves also a non-photochemical reaction, the so-called 'dark reaction'. No attempt will be made in the present account to consider the entire process, or even such parts of the entire process as have been elucidated for the Rhodophyta. For detailed accounts of the photosynthetic process in red and other algae the reader is referred to the standard works (Rabinowitsch, 1945, 1951, 1956; Haxo, 1960; Brody and Brody, 1962). The most important aspect of photosynthesis in red algae concerns the role of the accessory pigments and this alone will be treated in the present text.

Although photosynthesis in algae and higher plants is restricted to organisms which contain chlorophyll *a*, other pigments may make a contribution to the process. Engelmann (1883) devised an ingenious method in which the formation of oxygen was detected by means of motile bacteria. Using this, he concluded that pigments other than chlorophyll were effective in sensitising photosynthesis in brown, or red and blue-green algae. Engelmann showed that the red algae had the highest photosynthetic activity in the middle of the visible spectrum, an observation which is now well substantiated. Using more sophisticated procedures, it became possible to compare for any alga the absorption spectrum, that is, the percentage absorption of different wavelengths of light, with the relative photosynthetic rates, or action spectrum, measured by the output of oxygen. Superimposition of these two spectra demonstrates the relationship between light absorption and light utilisation at different wavelengths. For a green alga, such as *Ulva taeniata*, the correlation between the action and absorption spectra is extremely close (Blinks, 1954). In the case of the red alga *Delesseria decipiens*, the action spectrum does not correlate with the absorption spectrum for the thallus as a whole but only with the absorption spectrum for an aqueous extract, i.e. the phycobilin components. The very low photosynthetic rates for *Delesseria* at the 440 nm and 680 nm peaks for chlorophyll indicate that this pigment is relatively inactive. The energy absorbed by the phycoerythrin is passed, through phycocyanin, to chlorophyll where it then participates in the normal photosynthetic system. Providing that the accessory pigment is physically close to the chlorophyll the energy transfer can take place very rapidly, measurements of 0.5×10^{-9} seconds having been obtained for the transfer of energy to chlorophyll after its absorption by phycoerythrin.

A further conclusion from Engelmann's experiments was that algae synthesise best in light of the complementary colour. This topic has been investigated repeatedly (Fritsch, 1945; Biebl, 1962), both with respect to the colour changes of algae in different environmental circumstances and as an explanation for the zonation of marine algae in the intertidal and subtidal regions. Although both are aspects of the same problem, it might be best to look at each separately as they have been confused by some previous authors. It had been shown by the beginning of the present century that certain blue-green algae showed remarkable changes of pigmentation under laboratory conditions following exposure to light of different wavelengths. Red algae can undergo marked changes in colour although this appears to be due more to the destruction of phycoerythrin by light rather than to the induction of any new pigments. This is particularly apparent with the freshwater Rhodophyta which in most cases are brilliant green in colour. Various species of *Lemanea* and *Batrachospermum* collected under conditions of low light intensity exhibit shades of pink, red, purple or even black. On the other hand, certain species of *Batrachospermum*, such as *B. vagum*, are found frequently in small pools at high altitudes, where light values are particularly high, and these appear never to change colour. The green colour of various marine algae in the intertidal region is also an indication of photodestruction of phycoerythrin and is certainly no indication of chromatic adaptation as has been claimed by some. Turning next to the relationship between pigmentation and the zonation of marine algae, the widely accepted theory of chromatic adaptation postulates that the broad distribution of green, brown and red algae at successively lower levels of the intertidal and subtidal regions is a consequence of the adaptation of these groups of algae to the different light spectra to which they are subjected. An opposing view (Oltmanns, 1892) held that the zonation was not due to the quality but to the quantity of light. The demonstration (Harder, 1923) that both wavelength and intensity of light were important reconciled these two opposing views and there is now a considerable body of data in support of this view (Biebl, 1962). There are several difficulties in applying the general theory of chromatic adaptation to marine inter-tidal zonation. The most serious of these is that one cannot generalise about *all* red algae occurring in the subtidal and *all* green algae occurring in the high intertidal because there are too many species which are very obvious exceptions to this principle. The two red algae *Porphyra* and *Bangia* occur about as high in the intertidal as any marine algae, while some green algae such as *Ulva* can occur at considerable depths. On the other hand, the illumination conditions in the deep subtidal are those to which red algae are particularly suited. Under these con-ditions, red light has been reduced by absorption and, particularly in coastal waters, the blue light diminished by dispersion so that the light which remains tends to be distributed in the middle of the spectrum, the region most suited for absorption by phycoerythrin. In general, it must be stated that although the quality of the light may have some determining effect on the distribution of marine algae, a large number of other ecological factors must also be taken into consideration.

Although most marine and freshwater biologists have been aware of the role of algae as carbon-fixing organisms, it is only recently that there has been any appreciation of the need for quantitative measurement of the rates of organic production. The term 'productivity' is applied to the *quantity of organic material produced in a given time period.* This is *not* the same as 'standing crop' which is the *quantity present per unit area at a given instant,* although the two measurements have often been confused. The earliest calculations of primary productivity were derived from measurements of the long-term changes in standing crop and these were understandably somewhat uncertain. This procedure was soon superseded by the measurement of photosynthesis rates based on changes in gas concentrations in 'light and dark bottle' experiments. The latter are adequate for areas of high productivity but not for situations of low productivity where it is necessary to use an isotopic carbon technique. In most aquatic habitats, attention has been concentrated on the primary productivity of the plankton rather than the benthos and measurements of the productivity of benthic algae are relatively few, particularly for Rhodophyta. It has been shown that for red algae, the net production is of the order of 11 to 54 g dry weight per square metre per day (Blinks, 1955), giving values for carbon fixed (Pomeroy, 1961) as follows:

Iridaea sp.	2 g $C/m^2/day$
Gigartina sp.	7 g $C/m^2/day$
Porphyra sp.	1-3 g $C/m^2/day$

Methods for estimating productivity of benthic algae are derived from the methods developed for the phytoplankton and there is no one procedure which can be recommended for all situations. It is necessary to decide which method is best for a given purpose, modify it as required but then evaluate the results critically in terms of accuracy. It is unfortunate that the productivity of benthic algae should have been so neglected because in shallow waters it is not uncommon for this to exceed the productivity of the plankton. Coral reefs are a particular situation where the primary productivity may be due largely to the coralline algae, although endozoic algae may also be important, while the standing crop of plankton and the planktonic productivity are both extremely low. Two species of *Porolithon*, *P. gardineri* and *P. onkodes*, are major components of the algal ridges and the heavily grazed seaward slopes of coral reefs in Hawaii and the south Pacific. Estimates of the maximum total productivity for the two species are similar, about 0·5 g C/m²/day. The net productivity for *P. onkodes*, 2·2 g C/m² of thallus/day, and for *P. gardineri*, 2·4 g C/m² of thallus/day, lie within the ranges given above for other red algae (Littler, 1972). Physical factors of the environment influence strongly the productivity of larger benthic algae, particularly in tropical waters (Doty, 1971). For larger algae, the size of the standing crop on tropical reef flats is controlled as much by erratic storm turbulence as by predictable seasonal factors, although this may not be the case with smaller algae which are influenced less by the additional turbulence and more by seasonal factors. For the larger tropical algae, growth necessitates enhancement of diffusion rates either by storms or by tidal action. For this reason, it is essential that any conclusions regarding productivity derived from laboratory-based studies are confirmed in the field.

The whole of the previous discussion on primary productivity assumes that carbon is taken into the plant only in the form of CO_2 and, through the process of photosynthesis, is converted into complex organic forms. Several investigators have questioned whether benthic algae might not be dependent on processes other than photosynthesis for organic materials. There have been two principal marine ecological situations for which such suggestions have been made, arctic seas and the vicinity of outfalls discharging domestic sewage. Under arctic conditions, investigators have remarked frequently, since the initial comments by Kjellman (1883), that the sublittoral vegetation reaches its maximum development at a lower depth in the arctic than in the boreal. This is an area where there is no sun at all for several months of the year and where persistent ice covered by snow and slush is often present even when light has increased both in intensity and duration. The effect of this ice and snow is to diminish light intensity considerably, even at mid-day, and it has been reported that the arctic sublittoral at 100 m is virtually lightless even though large numbers of algae occur there, including several red algae. Crustose coralline algae, non-calcified crusts, and species with foliose thalli such as *Turnerella pennyi* (Wilce, 1967) have all been reported in such situations. Another instance which possibly is indicative of non-photosynthetic absorption of organic materials by marine algae relates to those species which occur in quantity in the vicinity of sewage outfalls where the species diversity is generally reduced. Some examples of these algae may represent nothing more than the ability to survive in adversity, as in the case of *Prionitis lanceolata* (Abbott, unpublished) which is the species last to be eliminated in the vicinity of outfalls in California. However, the

B

preponderance of coralline algae in such situations in many parts of the world has led to suggestions that they might be capable of non-photosynthetic heterotrophic existence. In freshwater habitats, a similar remark has been made about species which appear to persist in conditions of total darkness although there are few data on red algae under such circumstances, one of the few examples being *Audouinella chalybea* (Claus, 1961).

Heterotrophic abilities of algae in the laboratory have been examined by many investigators (Danforth, 1962) although much of the data cannot be applied to natural populations. Cell concentrations in such laboratory studies are far higher than those found in the field while the concentrations of organic materials which have been applied are also much higher than those found in natural waters. It has been argued (North *et al.*, 1972) that certain requirements must be met before the uptake of organic materials in natural populations can be accepted as a source of nutrient supply. First, the alga must be able to accumulate organic materials from very dilute solutions because the organic content rarely exceeds a few ppm even in heavily contaminated waters. Second, the alga must be shown to incorporate the absorbed material into its normal metabolic pathways. Finally, the organic material absorbed must be of significant proportions relative to that material produced from CO_2 through photosynthesis. Exposure of various red algae to dilute solutions of glycine shows that measurable accumulation and assimilation occur in certain cases (*Endocladia, Corallina, Gelidium* and *Lithothrix*) although many of the algae with fleshy thalli that were tested had little or no uptake ability. It is interesting to note that the species with uptake and assimilation abilities are those which do occur in waters receiving domestic sewage while those with little or no uptake ability are conspicuously absent. These are some of the first real data regarding heterotrophic ability in red algae and it is obvious that further examination is necessary. It is critical that the requirements listed above are applied if the data are to be of any significance in terms of the biology of the alga in the field.

Reserve materials and storage products

The characteristic reserve storage material found in red algae is floridean starch, first described by Kützing in 1843. Floridean starch occurs in grains, often of small size (*c.* 1 μm) although larger grains (ranging in size to as much as 25 μm) have also been recorded. They appear to lie free in the cytoplasm and although grains of floridean starch may be formed in proximity to the chloroplast they are never formed within the latter. When pyrenoids occur, floridean starch grains are deposited in their vicinity and they frequently occur adjacent to the nucleus. On treatment with iodine solution, the grains of floridean starch are stained yellow or brown, although a blue coloration may appear with prolonged exposure. Floridean starch is an α 1 : 4 glucan identical, or almost so, with the branched or amylopectin fraction of higher plant starches (Meeuse, 1962). The basic chain of floridean starch has an average of 15 glucose residues. One difference between floridean starch and higher plant amylopectins is that floridean starch gelatinises in water only after prolonged boiling.

Various other storage products (Meeuse, 1962) have been extracted from red algae. Trehalose has been detected in many freshwater Rhodophyta although its widespread occurrence in the marine representatives is somewhat doubtful. The 'trehalose' reported for *Rhodymenia palmata* (Kylin, 1915) has been reidentified as floridoside although it appears to be present, in minute amounts, in *Callithamnion tetricum*. Trehalose occurs also in the blue-green algae and has been reported for various groups of fungi. Floridoside, a glycerol galactoside, is extremely widespread in both Bangiophyceae and Florideophyceae and its concentration in *Rhodymenia palmata* may be as high as 15%. Isofloridoside has also been extracted from certain genera of the Bangiophyceae. Mannoglyceric acid is abundant in members of the Ceramiales, especially in genera of the Rhodomelaceae. In most cases where it occurs, the concentration of mannoglyceric acid is lower than that of floridoside, but this is not the case in *Ceramium rubrum*, while in *Griffithsia flosculosa* only mannoglyceric acid is reported. Sugar alcohols occur abundantly in many algae and various of these compounds have been reported from the red algae. Mannitol, long regarded as a characteristic storage product of the brown algae, not found in red algae (Fritsch, 1945), has been demonstrated in *Furcellaria*, *Halosaccion* and *Porphyra* so that it can no longer be regarded as a substance peculiar to the former group. Laminitol, which occurs in various brown algae, has also been reported from *Porphyra*, as well as volemitol, which is apparently absent from the brown algae. Sorbitol and dulcitol have long been known to occur in various red algae.

The nucleus

The interphase nucleus. The number of nuclei in each cell, their shape, size and arrangement in the cell vary considerably in the Rhodophyta.

In the Bangiophyceae, the cells are always uninucleate both in young and old tissues, even in the most elaborate thalli. The position in the Florideophyceae is more complex, the apical cells are either uninucleate or multinucleate, the segments formed by their division having at the moment of formation the same nuclear characteristics as the apical cell from which they have been produced. In some genera where the apical cell is uninucleate the cells of the mature thallus remain uninucleate throughout the life of the plant whereas in other genera there is an increase in the number of nuclei as a consequence of nuclear division without any corresponding cell division. It should be noted that in those genera where secondary pit connections are formed, nuclei are transferred between cells during the formation of these structures. Where the apical cells are multinucleate, these may contain a considerable number of nuclei and the number may continue to increase still further with age. The segments formed by the division of multinucleate apical cells are also multinucleate at the time of formation and the number of nuclei may increase as the cells mature. The largest numbers of nuclei are found in cells of various species of *Griffithsia*. In *G. globulifera*, it has been estimated (Lewis, 1909) that apical cells contain from 12 to 50 or even 75 nuclei with 100 to 500 nuclei in recently formed segments and 3000 to 4000 nuclei in the oldest cells, which are among the largest in the red algae. The numbers of nuclei in other red algae with multinucleate cells are still impressive, although not as high as those for *G. globulifera*; in *Pleonosporium borreri*, for instance, the apical cell may contain some 4 or 5 nuclei, with 20 to 30 in the older cells.

The position of the interphase nucleus or nuclei in the cell is related to the number of nuclei present and the state of the cell. In the Bangiophyceae, the single nucleus is placed centrally in young non-vacuolate cells and parietally in older vacuolate cells. Similarly in the Florideophyceae where a single nucleus is present in an apical cell or derivative cell this lies in a central position although after enlargement and the associated vacuolation the nucleus moves to the periphery of the cell. Occasionally, after vacuolation the single nucleus may be held in the centre of the cell by strands of cytoplasm but this is relatively rare. In genera where the thallus is obviously filamentous the single nucleus migrates to the basal pole of the cell where it lies in close proximity to the primary pit connection, as in *Ceramium*. Where the apical cell and its derivatives contain many nuclei, these lie at the periphery of the cell whether this is vacuolate or not.

According to Magne (1964a) the shape of the interphase nucleus is determined by its position in the cell and this hypothesis appears to be correct. Those nuclei lying in a central position are spherical whilst the parietal nuclei are somewhat flattened, with a lenticular shape. It has been claimed that nuclei with an irregular outline occur in certain genera and it is not clear to what extent these are simply fixation artefacts. Magne (1964a) has dismissed for this reason the statements of Svedelius (1911) that amoeboid nuclei occur in *Delesseria sanguinea* and this would appear to be justified in view of the fixative employed by the latter. On the other hand Magne also records nuclei with an irregular outline in the apical cells of *Gracilaria verrucosa* and in the apical cells of the filaments of limited growth in *Furcellaria fastigiata* and claims that these cannot be ascribed to bad fixation· There are also reports of the occurrence of elongate fusiform nuclei in certain tissues which may or may not be regarded as fixation artefacts. In old tissues where division of nuclei and cells has ceased the nuclei often exhibit curious changes in shape and appearance which appear to be a consequence of the disintegration of the nucleus. Clearly, further investigation of the phenomena are necessary before any adequate interpretation is possible.

The size of the nucleus shows considerable variation between species and also in different tissues of the same species, with further variation according to the time of year at which the material is examined and the conditions under which the plant is growing. The average size of the nuclei in the Rhodophyta is of the order of 3 μm. In general, the smallest nuclei are found in some genera of the Cryptonemiales and Gigartinales where the diameter may be less than 1 μm, whilst the largest occur in some members of the Ceramiales. The variation in size of the nuclei in a given plant is very complex as is shown by the detailed discussion of this given by Magne (1964a). The nuclei of apical cells, whether these are of filaments of limited or unlimited growth, tend to be larger than the nuclei of their derivative cells. The complications arise with older maturing cells where both cell division and nuclear division have ceased. Figures have been published showing nuclei of considerable size in such cells, as for instance in *Plocamium cartilagineum* where Kylin (1923, Fig. 34, D) figured a nucleus in face view with a diameter of 30 μm. The difficulty is that with vacuolation, the nuclei move to the periphery of the cell, becoming markedly flattened, so that accurate measurement of their dimensions is not easy, except in face view. However, if nuclear volume is calculated the increase in volume of the nucleus as the cell matures is considerable, far greater than could be accounted for by errors in measurement.

The early observations on the structure of the interphase nucleus are somewhat contradictory. The presence of one nucleolus, and occasionally more than one, was agreed by most investigators, although the presence or absence of a so-called 'chromatic reticulum' in the interphase nucleus was disputed. These differences of opinion were largely a result of inadequate technique and with the increased use of Feulgen staining the presence of fine strands in the interphase nucleus has been demonstrated.

Mitosis. From various studies of mitosis, in members of both Bangiophyceae and Florideophyceae, it would appear that the process is essentially similar to that found in higher plants (Dixon, 1966b). However, Magne (1964a, b) has claimed that in certain genera of the Gigartinales and Cryptonemiales there are certain important differences in both mitotic and meiotic nuclear division from that found in the majority of the Florideophyceae. The organisation in the former, which has been termed the 'calliblepharidian' nucleus by Magne, will be considered after a discussion of the more general type of mitosis found in the Florideophyceae.

The first indication of the commencement of mitosis is that the nuclei increase in size slightly. The subsequent changes are most easy to follow in those nuclei where fine 'chromatic' strands are readily observable in interphase. The chromocentres increase both in size and number and become more heavily stained. By the middle of prophase the fine strands representing the chromonemata have contracted so that the chromocentres have coalesced into chains of granules of varied length. In nuclei such as those of primary apical cells, where the interphase chromonemata are not readily observable as a result of the surface of the nucleus reacting as a homogenous Feulgen-positive region, the process is more difficult to follow but it appears to follow the same course. During late prophase the chains of chromocentres contract further to produce the chromosomes which lie at the periphery of the nucleus. The chromosomes at the end of prophase are of various shapes, some being elongate, others sausage-shaped, whilst the smallest are spherical. The shape of the chromosomes varies from species to species. Drew (1939) reported that the chromosomes of *Plumaria elegans* are of the three types listed above although she commented that this degree of variation is not found in other species that she examined.

In *Rhodochorton floridulum*, Magne (1964a) has shown that the chromosomes are all of the elongate type whilst in many other species, such as *Rhodymenia palmata*, the chromosomes are much smaller, being spherical or only slightly elongate. The one or more nucleoli also disintegrate towards the end of prophase although occasionally these may persist well into metaphase. Many of the early authors considered that some or all of the chromosomes were formed directly from the nucleolus. Westbrook (1935) reviewed these early observations and concluded that there was no real evidence to justify this opinion. Most recent work has supported the interpretation of chromosome formation given above. The nucleolus, at the time of disappearance, often fragments into a number of small diffusely staining structures and it may well be that as a result of the staining technique used these fragments were misinterpreted by the early workers.

During metaphase the chromosomes are usually contracted to the greatest extent. They move to the equator of the cell where they are very closely packed together. The aggregation of the chromosomes on the metaphase plate is so great that it is impossible to count the number of chromosomes present, even in polar view. The longitudinal partition of the chromosomes into chromatids is almost impossible to observe and only rarely has the occurrence of chromatids in metaphase been reported. This is not merely a result of their small size and close packing because they are still not visible when the cell is squashed so hard as to shatter the wall and spread the contents.

Critical observations on the spindle and polar bodies in the Rhodophyta are very few. Magne (1964a) has given a brief résumé of the previous observations and from a consideration of these and of his own investigation he concludes that in the 'normal' mitosis the mitotic figures are too small to permit any adequate analysis of the origin of the spindle. The outline of the spindle shows considerable variation in that the ends may be either pointed, truncated or round and it is peculiar that different types of spindle occur in different nuclei in the one plant. Polar bodies were described by all the earlier workers on red algal cytology but more recently the observations have been more critical. Westbrook (1935) was able to observe polar bodies in only two of the genera which she investigated and then only rarely, whilst more recently Magne (1964a) has commented that occasionally polar bodies could be observed but that there is considerable variation in appearance. For this reason, Magne argues that these polar bodies should not be regarded as centrosomes and this would appear to be a justifiable conclusion in view of the present inadequate knowledge. Until the reasons for their irregular appearance and ephemeral nature are understood it would be best simply to refer to these structures as 'polar bodies'.

The stages of anaphase and telophase proceed rapidly in all Rhodophyta. The two sets of chromosomes each tend to separate as a mass and little detail can be observed. Because of the close packing it is not possible to obtain any information on the centromatic attachment although, as will be seen later, some data on this feature can be obtained during meiosis. There has been some controversy as to whether the nuclear membrane breaks down or remains intact during mitosis in red algae. Reinvestigation (Magne, 1964a) of certain species where persistent envelopes had been reported failed to substantiate the earlier reports, but these differences of opinion are to be expected in light microscope studies of such extremely small nuclei.

As has been mentioned previously, Magne (1964a, b) has drawn attention to the occurrence in certain members of the Cryptonemiales and Gigartinales of differences in the process of mitosis. In these genera the chromosomes are formed in the normal way but the nucleus is markedly fusiform and in metaphase the polar bodies are very large and deeply staining. With the disappearance of the nuclear membrane the outline of the elongate nucleus remains, sharply circumscribed from the remainder of the cytoplasm. The polar bodies become more prominent during late metaphase and anaphase, becoming irregularly saucer-shaped.

Knowledge of the fine structural aspects of mitosis is extremely scanty, with but one published description (McDonald, 1972). This investigation showed that except for openings at the poles the nuclear envelope persisted throughout mitosis and provided much useful information on the structure described here previously as the polar body. Superficially the latter resembles a centriole in shape and diameter but it is shorter in length and lacks the 'cartwheel' substructure characteristic of the centriole in other plants. In the one species investigated, the body is associated with microtubules which have not been reported previously in red algae.

There appears to be considerable variation in the time at which mitosis occurs. Some investigators have claimed that mitotic figures could be obtained only at particular times or states of the tide in marine species, whilst other workers have found that mitosis occurred at all times. Even the same species may behave in different ways, in

Lemanea fluviatilis, for instance, mitotic figures were obtained at one site by day or night with no marked increase in frequency at any time, whereas at a second site, only a few miles from the first, mitosis could be observed only in material fixed between 10 p.m. and midnight. There have been some reports based on critical analyses which indicate the occurrence of mitotic periodicity in *Bangia fuscopurpurea* (Richardson, 1969) in the laboratory and in *Porphyra lanceolata* and *Rhodomela larix* (Austin and Pringle, 1968, 1969; Pringle and Austin, 1970) in the field. The last are probably the only studies of this phenomenon under natural conditions. On the other hand, the laboratory data for *Callithamnion roseum* (Konrad-Hawkins, 1964) gave no indication of any periodicity. It is not known if mitotic periodicity is of widespread occurrence and in those entities for which some evidence has been obtained it is not known whether this represents a circadian rhythm because the critical experimentation involving the continuation of the cycle under constant conditions has never been considered.

Meiosis. As with mitosis, the process of meiosis in the Rhodophyta is similar in most respects to that found in higher plants, although, as will be seen, there are some slight points of difference. As has been indicated by Magne (1964a), many of the early reports are incorrect in certain details. The various disagreements, particularly with regard to the relationship of the nucleolus to chromosome formation, are largely a consequence of faulty technique.

The early stages of the first prophase differ from the equivalent stages in mitosis in that the fine 'chromatic' strands (so-called 'reticulum') appear to be much more diffuse. Subsequently, a typical leptotene stage appears. Magne (1964a) has rightly drawn attention to the curious use of the term 'spireme' by both Drew (1934) and Westbrook (1935) to describe this highly filamentous state of the 'chromatin' in the Rhodophyta. The data on leptotene and zygotene stages are particularly sparse in Rhodophyta. The pairing of the chromosomes is often very difficult to interpret and it would appear that many of the observations on the phenomena described under the terms 'synapsis' or 'synizesis' are the result of aberrant fixation, as claimed by Austin (1960b). Magne (1964a), on the other hand, has figured various stages of zygotene which are very clear and he has shown also that in pachytene the chromosomes may not be fully paired along their whole length. The chromosomes, which up to this stage have the appearance of chains of granules, become thicker, shorter and more uniform in appearance. In the most favourable Feulgen preparations it is sometimes, but not always, possible to observe coiling, as indicated by Austin (1960b). During zygotene the chromosomes are frequently gathered to one side of the nucleus in such a way as to appear as a radiating mass. Further contraction occurs towards the end of pachytene and the paired chromosomes migrate again to occupy the whole of the nuclear area. In higher plants, as the diplotene stage proceeds, the paired chromosomes of the bivalent part and each chromosome is then seen to be composed of two chromatids. There is no real evidence for the occurrence of chromatids at this stage of meiosis in the nuclei of the Rhodophyta. This may be partly a consequence of the small size of the chromosomes but even in the larger nuclei, such as occur in the Ceramiales, chromatids cannot be seen. The crossed and looped arrangements of diplotene indicate the presence of one or more chiasmata. There is some variation in the number of chiasmata per bivalent, Austin (1960b) showing that on average there are two chiasmata per bivalent in *Furcellaria* whereas in some other genera there are fewer chiasmata. In favourable preparations it is usually possible to determine the position of the centromeres at this stage, whether lying in terminal, sub-terminal or median positions. One of the most characteristic features of meiosis in the Rhodophyta is that there is a marked interruption in the process towards the end of diplotene. The intensity of the staining reaction of the chromosomes falls off considerably so that they become much less obvious. This 'diffuse stage' which has been reported in various plants and animals (cf. Darlington, 1937) as well as in members of the Phaeophyta (Magne, 1964a) represents a partial reversion to the interphase condition. Throughout these stages of meiotic prophase the size of the nucleus has been increasing steadily and it reaches a maximum at the end of diplotene. During late diplotene the paired chromosomes shorten even more until at diakinesis they are represented by almost globular bivalents, although occasionally these may often have an angular appearance.

The nucleolus persists to the end of diakinesis. At this time it begins to disintegrate although occasionally the nucleolus may not disappear until well into metaphase. During metaphase the chromosomes become arranged on the equatorial plate, the nuclear membrane disappears and the volume occupied by the nucleus diminishes considerably. The arrangement of the spindle fibres is often clearer in the first meiotic metaphase than in mitosis, the fibres lying clearly between the polar body and the chromosome. In some nuclei the chromosomes have been described as consisting of numerous granules (Westbrook, 1928; Suneson, 1950), although in other nuclei massive hemispherical structures have been reported (Westbrook, 1935; Magne, 1964a). These structures appear during the diffuse stage of diplotene and have been identified as 'centrospheres' or 'centrosomes'. As Magne (1964a) has commented, until more knowledge is available on the fine structure of these objects it would be better simply to call them 'polar bodies'. Separation of the metaphase chromosomes take place rapidly once the orientation of the chromosome pairs on the equatorial plate has occurred. The anaphase and telophase stages of the first meiotic division do not exhibit any special features, the two sets of chromosomes each separating as a mass to the poles although occasionally an odd chromosome may lag behind its associates. The clusters of chromosomes, which form cup-shaped masses, despiralise and the individual chromosomes increase in length. The nucleoli and the nuclear membrane are reconstituted whilst the staining reaction of the chromosomes diminishes and an interphase intervenes. Yamanouchi (1906) alone has claimed that there is no interphase between the first and second divisions in the meiosis in *Polysiphonia violacea* (= *P. flexicaulis*), the two sets of chromosomes from the first division passing directly from telophase to the second division. Even in other species of the genus (Iyengar and Balakrishnan, 1949; Magne, 1964a) no evidence to support this view has been obtained. The second meiotic division does not appear to differ in any way from a normal mitosis except that the nucleolus and the nuclear membrane tend to disappear at a relatively early stage. The only calculation of the duration of meiosis appears to be the rough estimate given by Austin (1960b), who has stated that in *Furcellaria fastigiata* 'meiosis . . . takes a number of hours to reach completion, perhaps twelve or more'.

There has been no complete description of the fine structural aspects of meiosis in red algae. However, the one ultrastructural study (Kugrens and West, 1972) has shown that synaptonemal complexes occur in young tetrasporangia of at least four species of red algae. The synaptonemal complexes are similar in form and size to those found in other groups of plants where their occurrence is accepted as conclusive evidence for meiosis. Prior to this investigation they had not been reported in any algal group. The complexes were demonstrated in four species where meiosis was expected to occur but not in the young tetrasporangia of *Rhodochorton concrescens* which are thought to be apomeiotic (West, 1970). Providing that this relationship can be justified, the occurrence of synaptonemal complexes could be used to establish the site of meiosis in cases where this could not be demonstrated by classical cytological procedures using light microscopy. A second promising procedure for the same purpose involves the photometric measurement of DNA and it has been used to establish the position of meiosis in a species of *Batrachospermum* (Hurdelbrink and Schwantes, 1972).

The positions and times at which meiosis occurs are discussed elsewhere in connection with reproductive structures (p. 123) and life histories (p. 181).

Specialised cells and structures

Secretory cells

Secretory cells (or 'vesicular cells') occur in many genera of the Florideophyceae, representing all orders. Of the Nemaliales, secretory cells occur only in genera of the Bonnemaisoniaceae although in Cryptonemiales they occur extensively in genera of the Rhizophyllidaceae (*Rhizophyllis, Desmia, Ochtodes*), Cryptonemiaceae (*Halymenia*) and Peyssoneliaceae (*Contarinia, Peyssonelia*). Secretory cells have been reported in the Cruoriaceae (*Cruoria*), Nemastomataceae (*Nemastoma, Schizymenia*) and Solieriaceae (*Turnerella, Opuntiella*) of the Gigartinales. They are most abundant in genera of the Rhodymeniales, occurring in every genus of the Champiaceae and many of the Rhodymeniaceae (*Chrysymenia, Coelarthrum, Botryocladia, Coelothrix*) while in the Ceramiales secretory cells have been reported in a few genera of the Ceramiaceae (*Antithamnion, Ceramium, Dohrniella*) (see Fig. 1).

Where the thallus is relatively simple or obviously filamentous, as in genera of the Bonnemaisoniaceae or Ceramiaceae, the secretory cells are located in a lateral position on a filament. In more compact thalli, secretory cells usually occur in the outermost tissues where they are often formed by the transformation of the apical cell of a filament of limited growth, although in genera of the Rhodymeniales and a few other cases they are formed internally.

Although all are referred to as 'secretory cells' there is considerable diversity in their chemical composition although only a few have been investigated in any detail. The secretory cells at maturity contain little cytoplasm with no evidence of chloroplasts or photosynthetic pigments while the large and prominent vacuoles possess highly refractive contents. In the early stages of development, a prominent nucleus and a few chloroplasts may be present, although these disappear during

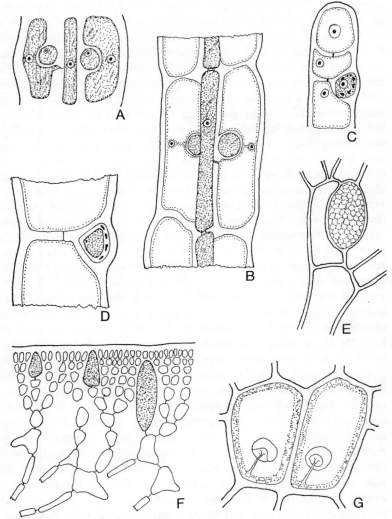

Fig. 1. *Secretory cells.* A–B, Falkenbergia *phase of* Asparagopsis armata; A, *development of secretory cell* (×1275); B, *mature aspect* (×700); C–D, Trailliella *phase of* Bonnemaisonia hamifera; C, *development of secretory cell* (×875); D, *mature aspect* (×950); E, Antithamnion elegans (×700); F, Schizymenia pacifica (×400); G, Laurencia obtusa (×525)

(A–D, F, *after Kylin;* E, G, *after Feldmann*)

maturation. The secretory cells in genera of the Bonnemaisoniaceae
are extremely prominent and they appear to be associated with high
concentrations of iodine. The concentration of iodine in such plants
may be sufficient for herbarium preparations to produce a strong blue
coloration with herbarium paper in which starch or a starch product
has been used as the 'filler'. The chemical state of the iodine was the
subject of much controversy in the early days of the present century,
some authors claiming that free iodine was present while others insisting
that it was merely an iodine-rich compound from which free iodine
was readily liberated (Kylin, 1928b). The latter reaction has been
attributed to an enzyme which has been termed 'iodine oxidase',
although it is highly probable that the 'iodine oxidases' of red and brown
algae are not identical in view of their very different properties. In
red algae, the enzyme to which this term is applied is readily soluble
and can be concentrated by precipitation with alcohol or ammonium
sulphate. The reactions are somewhat complex in that it has been
shown (Wolk, 1968) that in genera of the Bonnemaisoniaceae the
refractive contents characteristic of secretory cells were not formed
when the algae were cultured in the absence of bromide, although
vestigial secretory cells were still formed if the concentration was
reduced to one tenth of that found in natural seawater. The additional
observation that iodide oxidation is dependent on the presence of
bromide is particularly interesting in view of the structures which
occur in certain genera of the Ceramiaceae.

Secretory cells in which neither free iodine nor iodides can be
detected have been reported in various members of the Ceramiaceae,
such as *Antithamnion* and certain species of *Ceramium*. It has been
suggested (Sauvageau, 1926) that free bromine occurs in the secretory
cells of *Antithamnion* on the basis of the discovery that the secretory
cells in various species can convert fluorescein to its brominated derivative, eosin. This suggestion must be regarded with suspicion as the
cells are not capable of liberating free iodine from an iodide solution
as they would if free bromine were present. It is quite probable that
the situation is similar to that where it was argued whether free iodine
was or was not present in secretory cells.

In most of the Florideophyceae, the nature of the material secreted
is completely unknown. The secretory cells of *Peyssonelia rubra* appear
to contain crystalline material suggesting that they should be regarded
as equivalent to 'cystoliths', although those of the Rhodymeniales show
no identifiable product of secretion. It has been suggested that these

are responsible for the production of mucilage and a similar suggestion has been made for those found in *Halymenia*. In genera such as *Schizymenia*, *Turnerella*, *Opuntiella*, *Rhizophyllis* and *Ochtodes* the secretory cells are much larger than those discussed previously and are frequently referred to as 'gland cells'. In *Schizymenia*, the gland cells have homogenous contents and appear as brilliant points in face view of the thallus. They do not contain either iodine or iodides, they do not react with fluorescein as do the vesicular cells of *Antithamnion*, and with cresyl blue the vesicular cells show a turquoise coloration. The vesicular cells of *Turnerella* are very similar to those of *Schizymenia* but they occur deeper within the thallus rather than superficially. Of these large vesicular cells, those of *Ochtodes* are associated with the iridescent aspect of the thallus although this phenomenon does not appear to have been reported in any other examples.

Another type of vesicular arrangement is met with in *Laurencia obtusa* and other species of that genus. These are most obvious in superficial cells of the thallus and in the trichoblasts. Each cell contains a spherical body with a marked depression, so that it is reniform in section, which occurs at the tip of a prolongation of the peripheral cytoplasm and thereby lies in the vacuole. The spherical bodies possess extremely refringent contents in the living state and stain intensely with cresyl blue. Changes of considerable magnitude occur instantly as the cell dies but these bodies, referred to by French workers as *corps en cerise* (Feldmann and Feldmann, 1950) are of unknown function. The fine structure of the *corps en cerise* is totally different from that of any other organelle in the Rhodophyta (Bodard, 1968) and consists of a trabeculate structure containing grains of floridean starch and lipid globules.

The function of secretory cells is unclear in all cases. Iodine or iodine compounds certainly occur in some cases and there the formation is mediated by bromides. The evidence regarding the occurrence only of bromine or bromides in vesicular cells in the absence of iodine or iodides is less satisfactory and it should be appreciated that both iodides and bromides can occur in quantity in many red algae where vesicular cells have not been observed. Furthermore, there are the many examples of secretory cells which appear to have no connection with halide metabolism and are of completely unknown function.

Iridescent bodies and iridescence

Thalli of certain red algae show a marked blue or green iridescence when observed in reflected light, very different from the red coloration seen in transmitted light. The phenomenon appears to be most common in genera of the two orders Ceramiales and Rhodymeniales, although a few examples are also known in the Gigartinales and Nemaliales. Studies using optical microscopy indicated that several different structures were thought to be responsible for this phenomenon (Mangenot, 1933; Feldmann, 1937; Dangeard, 1940) although in certain cases there was no apparent morphological cause for the iridescence.

The most complete investigation is that by Feldmann (1937) who showed that the iridescent cells were of four types, of differing aspect, occurrence, content and chemical composition, as follows (see Fig. 2):

1. *Callithamnion* type. This type is most obvious in *C. caudatum* but it has been reported also in *C. tetragonum*, *C. granulatum* and in a species of *Seirospora*. In *Callithamnion caudatum*, spherical iridescent bodies often occur in pairs whereas in *Seirospora* they are of a more irregular elongate outline and usually occur singly in each cell. They appear to be found in all cells except the rhizoids and in the young condition have a homogeneous, finely granular structure. With cresyl blue, the vacuole is stained violet while the iridescent bodies are blue although osmic acid is not reduced by the latter. Histochemical tests suggest that the iridescent bodies are composed of proteinaceous material.

2. *Laurencia* type. This type has been reported in *L. pinnatifida* and *L. paniculata* and it is probably the type found in *Chondria coerulescens* and *C. scintillans*. The *Laurencia* type occurs only in some of the outermost cells of the thallus, which are usually slightly smaller than the cells of the thallus which do not contain iridescent bodies. The latter contain a prominent central vacuole whereas in the iridescent cells the vacuoles are filled with spherical bodies. With cresyl blue it is not possible to distinguish between the vacuole and the iridescent bodies in that both are coloured violet. Unlike the iridescent bodies of the *Callithamnion* type, those of the *Laurencia* type reduce osmic acid. They are clearly not of a proteinaceous nature and it has been suggested that they are tannin-rich or phenolic in composition.

3. *Gastroclonium* type. This type occurs in various members of the Champiaceae. The iridescent cells contain a large body which is often obviously laminated, located at the distal end of the cell. It does not stain with cresyl blue or neutral red and no attempt at chemical interpretation has been suggested.

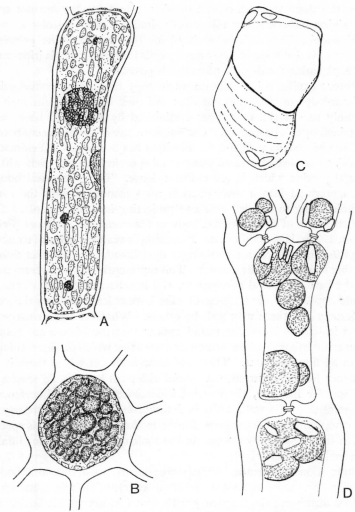

Fig. 2. *Iridescent bodies.* A, Callithamnion caudatum (×700); B, Laurencia paniculata (×700); C, Chyclocladia clavata (×880); D, Chondria coerulescens (×760)

(A, B, *after Feldmann;* C, D, *after Mangenot*)

4. *Ochtodes* type. The iridescent cells of this alga differ completely from the three types discussed previously in that they are not cells containing inclusions but rather cells similar to the vesicular cells of *Schizymenia*, *Turnerella* or *Rhizophyllis* (see p. 33). The principal difference is that none of the secretory cells in those genera is iridescent. No explanation for this phenomenon is possible.

Recent studies using electron microscopy have added considerably to knowledge of the iridescent cells and their inclusions, although so far only two of the four types established by Feldmann have been examined by this procedure. The fine structure of the iridescent bodies in *Chondria coerulescens* and *C. scintillans* has been shown (Feldmann, 1970a) to consist of a large number of spherical bodies, each with a central portion which is not electron dense. These spherical bodies are aggregated into a hemispherical mass associated with the cytoplasmic lining to the cell wall and projecting into the vacuole. The fine structure of the iridescent body in *Gastroclonium clavatum* (Feldmann, 1970b) is very different. It consists of regular chains of elongate units 0·1 μm in breadth and longer than broad, lying parallel to one another in a very regular manner. This represents what was interpreted, on the basis of optical microscopy, as a lamellate structure. Thus, it would appear that the iridescent bodies have at least two very different structures and there may well be others. Whatever the structure of these iridescent bodies, it would appear that the iridescence results from an interference phenomenon similar to that which produces colours in an oil film on water. There are some instances where there is no evidence for the occurrence of special iridescent bodies. In species of *Iridaea*, for instance, the marked iridescence is due to interference phenomena in the wall, while in *Fauchea laciniata* some property of the cytoplasm itself is the cause for the marked blue cast of this species which is modifiable by changes in the tonicity of the medium (Blinks, personal communication).

Suggestions (Oltmanns, 1905) that the occurrence of iridescent bodies provided protection against the injurious effects of high light intensities can be dismissed; the frequency with which iridescent Mediterranean species occur only in grottoes (Feldmann, 1937) and the particularly strong blue cast of *Fauchea laciniata* in low light (Dixon, unpublished observations) are sufficient to discount this hypothesis which was based on speculation devoid of experimental evidence. It is impossible to state, at the present time, whether iridescence has any purpose or function.

Hairs (see Fig. 3, A-D)

The hairs which occur in many Florideophyceae are formed from superficial cells of the thallus in most cases, that is, from the apical cells of the filaments of limited growth. In various species of *Melobesia*, the hairs termed 'heterocysts' by Rosanoff (1866) and 'trichocytes' by Rosenvinge (1917) are formed as lateral structures on the cells of the prostrate axes. It is said that such cells lack the cap-cells (see p. 57) but this is not always the case. The first step in the formation of a hair, a bulbous protuberance forming on the outer face, is very similar to the equivalent step in the formation of a carpogonium so that there have been numerous instances of hairs being misidentified as carpogonia, particularly by the earlier workers. The protuberance elongates apparently at some speed, although no quantitative data are available. The enlarging hair is a cylindrical structure, with a large central vacuole surrounded by a thin layer of cytoplasm which lines the wall. The tip of the hair is not vacuolate and a nucleus can sometimes be detected. As growth continues, the amount of cytoplasm at the tip gradually diminishes to a mere meniscus, by which time the nucleus has disappeared. Chloroplasts are present in the cell from which the hair originates and may occasionally occur in the hair itself. Hairs are ephemeral structures, although there is no exact information as to their longevity. As with carpogonia, a knowledge of the fine structure of the hair in relation to its development would be extremely useful, particularly with regard to the resolution of the many questions regarding its function (Fritsch, 1945).

The occurrence of hairs shows various peculiarities, in that they appear to be totally lacking in members of the Bangiophyceae while in Florideophyceae they appear never to have been detected in members of the Gigartinaceae and Phyllophoraceae (Gigartinales) or Delesseriaceae, Dasyaceae and Rhodomelaceae (Ceramiales). Furthermore, the frequency of occurrence of hairs in species for which they have been reported varies enormously from specimen to specimen and fluctuates with season and environmental conditions. In algae of the northern hemisphere in which hairs occur, they are often said to be most abundant in spring and early summer but absent during the winter months. This could well be a reflection merely of their formation during active thallus growth and of their essentially ephemeral nature. Nothing is known of the function of hairs. In view of their relative predominance in well-illuminated specimens or at the beginning of the growing season, it has been suggested that they serve as a protection against

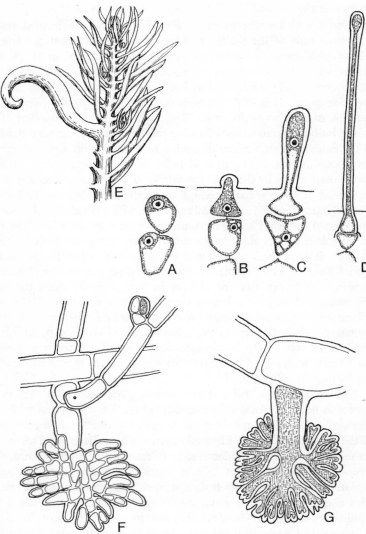

Fig. 3. *Specialised cells and structures.* A–D, Cystoclonium purpureum; A–C, *development of hair* (×900); D, *mature hair* (×430); D,
Bonnemaisonia hamifera, *attachment structure* (×10); F, Antithamnion cruciatum, *multicellular rhizoidal attachment* (×235); G, Spermothamnion flabellatum, *clasping rhizoid formed from a single cell* (×235)
A–D, *after Kylin;* E, *after Taylor;* F–G, *after Feldmann-Mazoyer)*

illumination damage (Berthold, 1882a) or as an absorbing surface for nutrients during active growth (Oltmanns, 1904; Rosenvinge, 1911; Kylin, 1917a), but there are no experimental data to support either of these hypotheses.

Rhizoids, tendrils and attachment structures (see Fig. 3, E-G)

Most thalli, of both the Florideophyceae and Bangiophyceae, are attached to the substrate on which they develop by rhizoidal structures of various forms.   In some cases, the thallus is attached directly to the substrate and rhizoids are lacking.   It would appear that this type of attachment represents some form of 'cementation' and this could be compared with the adhesions between filaments found in all complex thalli.   The rhizoids are essentially simple, single-celled structures throughout the Bangiophyceae, although they may be produced in such quantity in *Porphyra* to form a basal discoid attachment structure of some size.   Simple, single-celled rhizoids are produced by the majority of genera in the Florideophyceae.   In many Ceramiaceae, the tip of the rhizoid may remain simple or a several-celled hapteron-like clasping structure may form at the apex.   Most rhizoids do not penetrate the substrate, although the attachment of *Polysiphonia lanosa*, an obligate epiphyte on *Ascophyllum* and other fucoids, consists of a single swollen rhizoid which penetrates the tissue of the substrate species. Some thalli of the Florideophyceae possess a very different type of attachment structure resembling a mass of rhizoids adhering laterally to form a relatively massive multicellular structure.   Such massive attachment structures may be formed laterally on an axis, as in *Gelidium* or various genera of the Delesseriaceae, or terminally by the conversion of the apex of an axis, as happens frequently in *Plocamium*.   As far as can be ascertained, these multicellular rhizoidal attachments develop in response to a contact stimulus.   Whether formed laterally or terminally, the attachment of prostrate axes of various algae at intervals produces an organisation similar to the stolons of higher plants and there is evidence that such algae possess certain of the features of vegetative propagation found in stoloniferous Angiosperms (see p. 211).

Special curved laterals which, after contact with the same or another alga, or with some foreign material, can form one or more firm coils around it, after the manner of a tendril in the Angiosperms, have been reported in diverse Florideophyceae.   There are usually morphological differences in the degree of differentiation between the tissues of the inner and outer surfaces of the curved laterals prior to contact and coiling. The cells of the two surfaces may differ markedly in size and the formation of lateral branches tends to be restricted to the outer face. After contact, the smaller cells on the inner face enlarge to form attaching structures which may be either unicellular rhizoids or multicellular structures. The degree of incurvature of the laterals varies considerably from species to species. The curved laterals of *Bonnemaisonia hamifera* and *B. nootkana* rarely curve more than half a turn, whereas in *Cystoclonium* and *Calliblepharis lanceolata* the tendrils are particularly long twisted structures which may make as many as 4 to 5 coils around the object to which they are attached. It has been assumed, and there is a little experimental evidence to support the view, that the inrolling of the tendril and the production of attachment structures on the inner face are the result of a contact stimulus. What is not clear is the nature of the stimulus, which appears to function in different species in different ways. For instance, in *Cystoclonium*, the tendrils develop in quantity and attach particularly strongly when an axis of the same entity is enrolled whereas in other species the degree of enrolment is stimulated not at all by an axis of the same species but strongly by certain other materials. Where an axis is encircled by a tendril of the same species, tissue fusions can develop which are of a much more intimate nature than rhizoidal attachments.

The short reflexed laterals found on the spine-like branches of *Asparagopsis armata* do not possess any capacity to encircle objects with which they come in contact, although after a firm attachment through one of the hooks has developed the tip of the spine forms a terminal rhizoidal attachment structure similar to that of *Plocamium*. Many species which do not possess tendril-like laterals may undergo tissue fusions either with axes of the same or another species. As with the tendrils, attachments through rhizoids develop frequently when two axes of the same or different species come into contact. Sometimes rhizoids will develop from both participants, although it is said that such reciprocal rhizoid development only occurs when two axes of the same species are involved (Fritsch, 1945). In many cases, more intimate contact may develop, even to the extent of secondary pit connection formation, and there is some evidence to indicate that this can take place between axes of different species under certain circumstances.

Galls and tumour-like growths (see Fig. 4)

Gall-like proliferations occur on a wide variety of red algae (Merola, 1956; Tokida, 1958), although they appear to be very much more

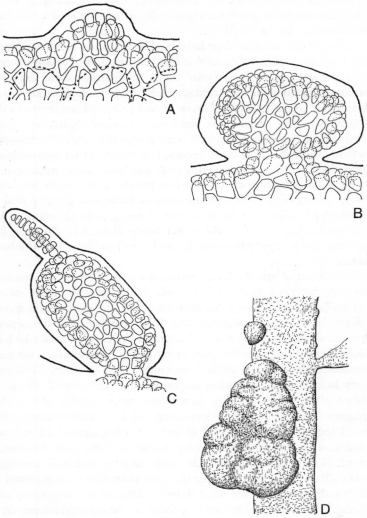

Fig. 4. *Galls and tumour-like growths.* A, Ceramium *sp., early stage of development of gall* (×270); B, Ceramium *sp., gall* (×225); C, Ceramium *sp., gall from which a new axis is differentiating* (×225); D, Prionitis lanceolata, *gall previously regarded as the parasite* Lobocolax deformans (×5)

(A, B, C, *after Dixon;* D, *after Smith*)

frequent on members of the Florideophyceae than Bangiophyceae. They develop as irregular masses of cells, of varying size, which may be discernible to the naked eye.

Such gall-like proliferations have been the cause of much confusion. The early descriptions of *Ceramium botryocarpum* and *C. deslongchampsii* were characterised by 'cystocarps' which were devoid of the cluster of enveloping involucral axes characteristic of the carposporophytes of this genus; these 'cystocarps' were in fact galls and not reproductive structures. The parasites or supposed 'parasites' which have been recorded on various Rhodophyta were frequently confused with galls so that *Choroeolax cystoclonii*, reported as occurring on *Cystoclonium*, and *Lobocolax deformans*, occurring on *Prionitis lanceolata*, would appear to be nothing more than tumour-like proliferations of the supposed hosts. Many of the reported instances of parasporangia may well be based on an indiscriminate, disorganised cell growth although, as has been shown (p. 133), it could be extremely difficult to draw the line between a cluster of parasporangia and a gall in terms of their basic morphology.

Various causal agencies have been shown to produce galls or tumour-like growths in red algae. The possible role of bacteria in this respect was initially suggested by Schmitz (1892) in the studies which first differentiated between such galls and those protuberances due to parasitic red algae. In some cases a bacterium has been isolated which on reinoculation caused development of new galls. Conversely, it has been demonstrated (Brucker, 1958; Künzenbach and Brucker, 1960) that many isolates of the organism responsible for 'crown-gall' disease in higher plants (*Agrobacterium tumefaciens*) were capable of inducing tumours similar to naturally occurring gall-like protuberances when inoculated into a large number of red and other algae. All galls are not due to primary bacterial activity in that there are several examples for which such an explanation is completely untenable (Rosenvinge, 1931). Both fungi and endophytic algae have also been reported as causing distortions in red algal tissues. Despite the frequency with which phycomycete and ascomycete fungi can be detected on red algae, the production of galls by those organisms is relatively rare. Endophytic algae by comparison cause much more distortion of growth patterns than fungi although it is unusual for such activity to produce a macroscopic gall. A species of *Chlorochytrium* causes considerable distortion in the frond of *Porphyra tenera* and similar distortions caused by a unicellular green alga possibly referable to the genus *Chlorochytrium* have been observed in various genera of the Delesseriaceae both in Europe and the Pacific coast of North America (Dixon, unpublished observations). Endophytic ectocarpaceous brown algae of the type usually attributed to the genus *Streblonema* cause distortions of various sorts in red algae. Such distortions are most readily observed in flat membranaceous thalli such as those of the Delesseriaceae although they have been observed also in *Gelidium, Pterocladia, Furcellaria* and other red algae of more cartilaginous construction. Usually cells in the immediate vicinity of the infection are killed while cells at some distance begin to divide in an uncontrolled manner and this produces a gall or tumour. Clusters of diatoms which develop internally may also give rise to tissue distortions.

As with higher plants, algal galls may be produced as a consequence of animal infestation and such galls occur frequently in the red algae. In general two types of animal are involved—nematodes and harpacticoid Copepods. Nematodes have been detected in galls of *Furcellaria fastigiata* and *Chondrus crispus* and the organism is similar to that nematode which produces galls in *Ascophyllum nodosum* and other fucoid algae. Harpacticoid Copepods infest red algae, producing disturbances and galls much more frequently than nematodes. Early reports (Barton, 1891; Brady, 1894) must be treated with caution although more recent studies (Bocquet, 1953; Harding, 1954) give detailed accounts of the biology of the infestants in respect to the behaviour of the host alga. Such Copepods have been reported in *Rhodymenia palmata, R. pseudopalmata, Rhodophyllis divaricata, Stenogramme interrupta, Polyneura gmelinii, Cryptopleura ramosa* and *Nitophyllum bonnemaisonii* and probably occur even more extensively.

Finally, it must be indicated that some galls or tumours in red algae are undoubtedly caused by chemical factors in the environment. The large scale culture of *Porphyra* in Japan (see p. 217) has disclosed the frequent occurrence of galls in the vicinity of sewage discharges with a high proportion of industrial wastes (Fujiyama, 1957; Katayama and Fujiyama, 1957). Investigation shows that these industrial wastes appear to produce bi- and tri-nucleate cells which then become the sites of origin of abnormal growths. It could well be that many of the reports of galls where no obvious cause can be detected may be produced in this way. Virus infection has also been suggested although there is no evidence to support this at the present time.

Under certain circumstances, irrespective of the nature of the causal agent, the gall may give rise, from this mass of disorganised cells, to new axes of apparently normal structure and organisation.   Such reversion occurs very frequently in the galls of *Cystoclonium purpureum* and has been detected on several occasions in species of *Ceramium* (Dixon, 1960).

The way by which galls or tumours are formed in red algae, the factors responsible for their development and possible reversion are areas about which the level of knowledge is very low.   Lewin (1962) has indicated that experimental investigation of galls has been largely ignored.   While this is true, the basic reason for this neglect is that the average phycologist would probably not even be able to recognise a gall, so great is the ignorance of these structures.

2
Morphology

There are considerable variations in the level of organisation and degree of complexity of the thalli throughout the Rhodophyta. With the possible exception of *Compsopogon*, the thalli of genera of the Bangiophyceae are relatively simple and include even several unicellular representatives. Thalli of members of the Florideophyceae, on the other hand, are much more complex and elaborate. Without exception, patterns of symmetry are present involving control of cell division and enlargement, apical dominance and interaction between filaments of different orders.

Morphology of the Bangiophyceae

Several genera of the Bangiophyceae consist only of unicells or the irregular masses of cells which result from the products of the division of single cells failing to operate. The best known of these genera is *Porphyridium* (Fig. 5, c). One species, *P. sanguineum* (frequently referred to in the literature as *P. cruentum*) occurs on soil or damp walls where it forms blood-red mucilaginous masses. Several species of this genus have also been described from the marine plankton. The individual cells of *Porphyridium* are globose, each with a prominent stellate chloroplast and central pyrenoid. The other genera in which the thallus consists of a single cell are distinguished from *Porphyridium* largely on the basis of cell shape, the thickness of the cell wall, the position of the chloroplast and the presence or absence of a pyrenoid (Kylin, 1956). Most of these genera of the Bangiophyceae in which the thallus is unicellular are virtually unknown and many have been described from extremely curious habitats. *Vanhoffenia* is known as an epiphyte on *Nitella* and moss leaves from Kerguelen Island while the

45

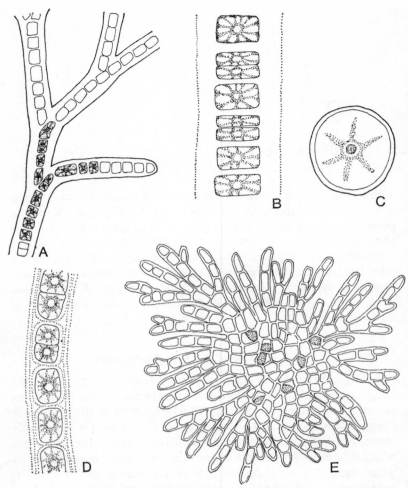

Fig. 5. *Simple thallus morphology in some genera of Bangiophyceae.*
A, Goniotrichum elegans (×350); B, Goniotrichum elegans, *portion showing cell structure* (×800); C, Porphyridium marinum (×2100); D, Asterocytis ramosa (×900); E, Erythrocladia irregularis (×700); F, Goniotrichopsis sublittoralis, *young plant* (×350); G, Goniotrichopsis sublittoralis, *older plant* (×350); H, Goniotrichopsis sublittoralis, *portion showing cell structure* (×950); I, Erythrotrichia carnea, *young plant* (×950)

(A, *after Rosenvinge;* B, D–I, *after Smith;* C, *after Kylin*)

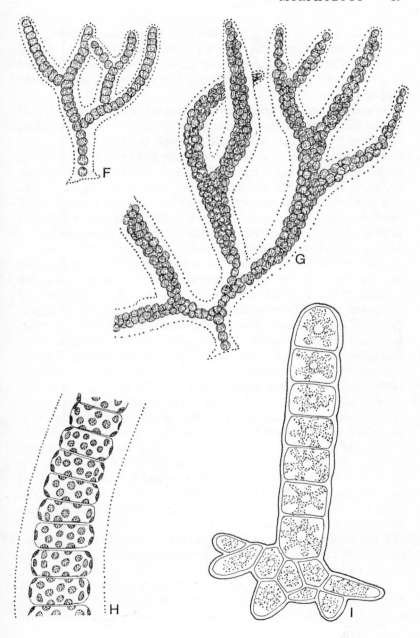

genera *Chroothece* and *Petrovanella* were described from saline soils in central Europe. It is very probable that their relatively infrequent detection should not be taken as a real indication of rarity. Those genera in which the thallus is either a unicell or the colonial product of a unicell are grouped in the family Porphyridiaceae of the order Porphyridiales.

By comparison, the genera of the Goniotrichaceae possess multicellular thalli in which there is some degree of organised cell orientation. Species of *Goniotrichum* have small epiphytic thalli (see Fig. 5, A, B), with branched filaments which remain uniseriate in some (*G. elegans*), but become multiseriate in others. Branching of the uniseriate filament can occur in a manner analogous to the 'false branching' of certain blue-green algae. The monotypic *Goniotrichopsis* of the Pacific coast of North America is indistinguishable from *Goniotrichum* morphologically, although differing markedly in that each cell contains several peripheral discoid chloroplasts devoid of pyrenoids, rather than the single parietal chloroplast with central pyrenoid found in every cell of *Goniotrichum* (see Fig. 5). The exact differences between the thallus of a genus assigned to the Goniotrichaceae and a colonial cell mass formed by a member of the Porphyridiaceae are extremely difficult to define. Species of *Asterocytis* may form a thallus consisting of cells placed end to end (see Fig. 5, D) although they may also form irregularly branched multiseriate cell masses. The individual cells are virtually identical with the unicellular *Chroothece* and previous reports (Belcher and Swale, 1960; Pujals, 1961) have commented frequently on their release. A recent study (Lewin and Robertson, 1971) has shown that an *Asterocytis* of marine origin changed its morphology from filamentous to a unicellular or bicellular form resembling *Chroothece* with reduction in the salinity of the seawater medium in which it was growing in pure culture. The effects of salinity cannot be the only factor involved in that *Chroothece* is reported from saline soils and freshwater *Asterocytis* has been described. Because of this *Chroothece/Asterocytis* relationship, it has been proposed (Feldmann, 1955) that the separation of the Goniotrichaceae as a distinct order, the Goniotrichales, is unwarranted and that the family should be merged with the Porphyridiales, an opinion accepted in the present systematic arrangement of the Bangiophyceae (see p. 241).

A much higher level of organisation is met with in thalli of genera assigned to the order Bangiales, the least specialised members of which are grouped in the Erythropeltidaceae. Species of *Erythrotrichia* are small epiphytes, in some of which the thallus is differentiated into a prostrate basal portion from which one or more erect fronds arise (see Fig. 5, I). The degree of development of the basal disc relative to that of the erect frond varies considerably. In some specimens, the thallus may consist of only or little more than the disc and this can give rise to reproductive structures. The genus *Erythropeltis* was established for *Erythrotrichia discigera*, a species in which this reduction of the thallus to the basal disc occurs extremely frequently. From what has been said, it is obvious that the validity of the genus *Erythropeltis* is open to severe question. Species of *Erythrocladia* form slightly more complex discs than those of *Erythropeltis* but their degree of distinctness is equally open to doubt. As will be seen, the thallus of *Porphyropsis* passes through a stage in development which is not distinguishable from that of a species currently assigned to *Erythrocladia*. The thallus of *Porphyropsis* is a flat sheet of cells, similar in some respects to certain species of *Porphyra*, but differing markedly in its development as well as reproduction and life history. The thallus of *Porphyropsis* develops from a discoid base into a saccate structure which eventually breaks up to give a minute sheet of cells, in a manner similar to the development of the thallus in the green alga *Monostroma*. Genera of the second family of the Bangiales, the Bangiaceae, have thalli which are much more complex than those of the Erythropeltidaceae and differing fundamentally in that there is no trace of a heterotrichous crustose base. The product of spore germination develops in an upright manner and is attached initially only by the basal cell. A substantial base can develop with time as a result of the secondary development of downgrowing rhizoidal cells from the lowermost cells of the thallus. In *Bangia*, the young thallus consists of a uniseriate row of cells and it may remain in this state throughout its life. It is more common for each cell of the uniseriate thallus to become divided, by a series of radially-arranged walls, to produce a multiseriate condition (see Fig. 6, C, D). The thalli of *Porphyra* also begin as uniseriate filaments but this stage is quickly passed and large parenchymatous sheets of cells, one or two cells in thickness, are produced as a result of abundant intercalary cell division (see Fig. 6, A, B). There have been suggestions that the growth of the uniseriate filament in both *Bangia* and *Porphyra* is a consequence of strictly apical division. While this might be the case when the filament is one or two cells in length in *Bangia*, the occurrence of intercalary cell division is by far the more significant contribution once the thallus has achieved a length of five or six cells. In *Porphyra* a similar situation

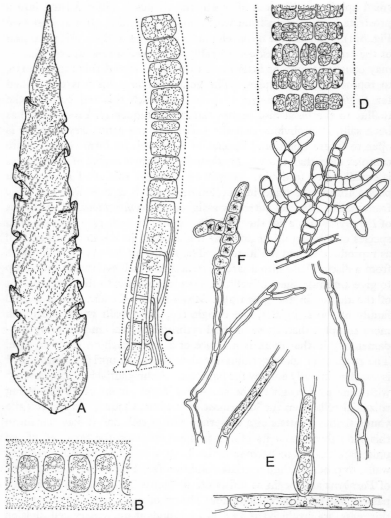

Fig. 6. *Morphology of Bangiophyceae.* A, *thallus of* Porphyra occidentalis (×0.5); B, *section of thallus*, Porphyra pulchra (×450); C, *uniseriate basal portion of thallus*, Bangia fuscopurpurea (×400); D, *upper multiseriate portion of thallus*, Bangia fuscopurpurea (×350); E, Conchocelis *phase of* Porphyra *sp.* (×380); F, Conchocelis *phase of* Porphyra *sp., with 'fertile cell rows'* (×170)

(A–D, *after Smith;* E–F, *after Kornmann*)

exists although here abundant intercalary cell division in two planes occurs in a plant of this size (see Fig. 6, A, B). The intercalary cell division in certain species of *Porphyra* can produce extremely rapid growth, thalli of *P. miniata* achieving a length in excess of one metre in less than three weeks growth in the field. Various species of *Porphyra* are either collected 'wild', or cultivated, as an item of diet (see p. 215).

The order Compsopogonales contains but one genus, *Compsopogon*, of which about a dozen species have been described from tropical and subtropical freshwaters (see Fig. 7, A, B, C). The thalli in this genus are by far the most complex of any member of the Bangiophyceae. Each axis has a prominent apical cell which divides transversely to give a filament of short discoid segments (Fig. 7, A). Branching of the thallus occurs at this stage of development by the formation of a lateral apical cell through an oblique division of a segment. In certain cases, development of the thallus does not appear to proceed further than the simple or branched uniseriate condition, the only change being that elongation of the segments occurs so that their outline changes from discoid to a more quadratic shape. Further segmentation occurs in most thalli producing a compact investment of small cells which cover the segments formed by the division of the apical cells. The investment may become more than one cell in thickness in certain cases (Fig. 7, B, C). The final product of thallus development in species of *Compsopogon* bears some analogies to the thalli of *Ceramium* (Florideophyceae, p. 75) although the pattern of development is very different.

The order Rhodochaetales also contains but one genus, *Rhodochaete* (Fig. 7, D-G). *Rhodochaete parvula* has been detected on only a very few occasions from several localities in the Mediterranean and West Indies. The thallus is composed of infrequently branched uniseriate filaments composed of elongate cells, each of which contains several elongate parietal chloroplasts. The general aspect of the plant is similar to those of certain species of the *Acrochaetium/Rhodochorton* complex and in view of the reported pit connections between adjacent cells, the attribution of this curious organism might well be questioned.

Morphology of the Florideophyceae

All thalli in the Florideophyceae are formed by the aggregation of filaments so that they are essentially of a pseudoparenchymatous nature. This interpretation has been challenged by Bugnon (1961), but an

c

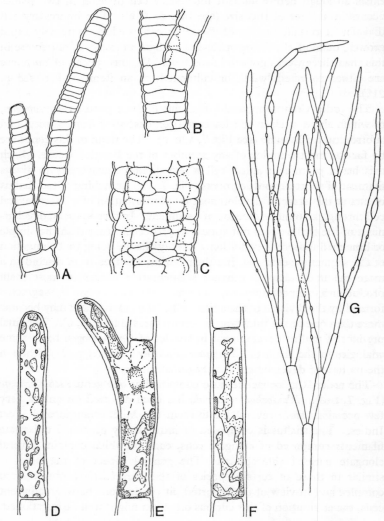

Fig. 7. *Morphology of Bangiophyceae.* A–C, Compsopogon coeruleus; A, *apex of thallus* (× 200); B–C, *stages in the development of multiseriate condition* (× 200); D–G, Rhodochaete parvula; D, *apex of filament* (× 1000); E, *cell of filament with origin of lateral filament* (× 1000);
F, *mature cell of filament* (× 1000); G, *erect frond* (× 270)
(A–C, *after Boillot;* D–G, *after Magne*)

alternative explanation for his arguments is presented later (see p. 111). The filaments comprising the thalli are usually formed by the division of apical cells, although intercalary cell division does also occur in certain genera of the Delesseriaceae and Corallinaceae. The possibility that intercalary cell division might occur in other Florideophyceae should not be neglected without further consideration, since a failure to detect intermittent intercalary cell division in thalli where its occurrence is not anticipated could lead to extremely serious errors of interpretation. At each division of an apical cell, or with the formation of a lateral apical cell, a pit connection is formed and it is possible therefore to deduce the pattern of cell division by which a thallus has arisen from the pattern of pit connections. The formation in some thalli of secondary pit connections between cells which are not of kindred origin, must of course be taken into consideration in such analyses.

The thalli of the Florideophyceae vary considerably in their structure, development and appearance (see Fig. 8). It has been customary (Oltmanns, 1904, 1905; Kylin, 1937a, 1956; Fritsch, 1945) to interpret thallus organisation in the Florideophyceae on the basis of two general categories depending upon whether there was one filament or a group of filaments forming the core in each axis. The former was referred to as 'uniaxial' or 'Zentralfadentypus', the latter as 'multiaxial' or 'Springbrunnentypus'. What does not appear to have been appreciated is that this scheme is applicable, strictly speaking, only to those thalli which are erect, 'foliose', structures and attempts to apply this concept in any interpretation of crustose thalli are totally irrelevant. From the morphogenetic point of view two basic types of construction can be distinguished in the Florideophyceae but these are not delimited on the basis of axial structure. The two types are the encrusting thallus and the erect foliose thallus, which may be interpreted as the consequences of elaboration of either the prostrate or the erect components, respectively, of a structure with heterotrichous organisation.

The encrusting thallus

In certain genera of the Gigartinales and Cryptonemiales, referred principally to the Cruoriaceae and Peyssoneliaceae (formerly the Squamariaceae) respectively, the plant has a relatively simple and, usually, very obvious heterotrichous organisation (see Fig. 9). It consists of an encrusting, circular or irregularly lobed, reddish, purple or blackish thallus. This consists of a basal layer of aggregated, radiating filaments produced by the germination of a spore (Killian, 1914). All cells of the

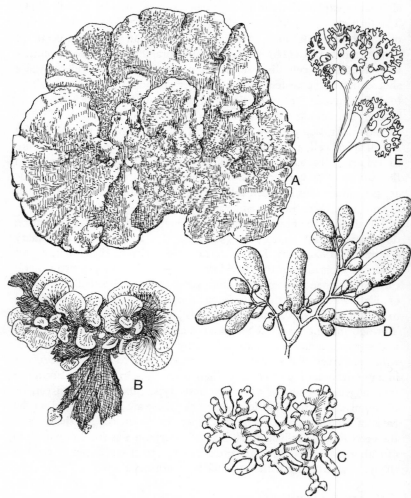

Fig. 8. *Form of the thallus in certain Florideophyceae.* A, Lithophyllum expansum (× 1); B, Rhizophyllis squamariae (× 1); C, Phymatolithon (Lithothamnium) calcareum (× 1); D, Botryocladia pseudodichotoma (× 0.5); E, Chondrus crispus (× 0.5); F, Batrachospermum moniliforme (× 4); G, Grinnellia americana (× 0.35); H, Ahnfeltia plicata (× 0.7); I, Corallina officinalis (× 3)

(A, B, *after Oltmanns;* C, *after Newton;* D, *after Smith;* E, G–I, *after Taylor;* F, *after Sirodot*)

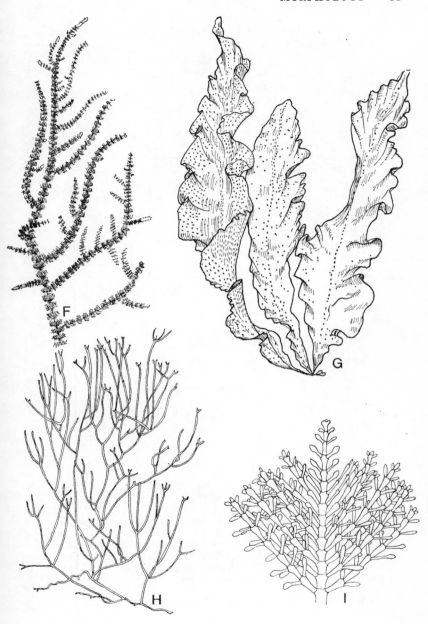

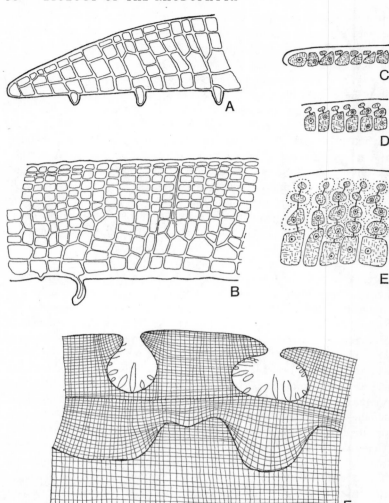

Fig. 9. *Crustose thalli based on a primary heterotrichous organisation.*
A–B, Peyssonelia dubyi; A, *section of thallus edge* (× 325); B, *section
of mature thallus* (× 250); C–E, Melobesia membranacea; C, *section of
thallus edge* (× 600); D, *section of older thallus* (× 600); E, *section of
mature thallus* (× 730); F, Hildenbrandia prototypus, *section showing
the annual increments and positions of conceptacles in previous year's
growths* (× 250)

(A, C–E, *after Kylin;* B, F, *after Rosenvinge*)

prostrate system give rise to upright filaments which are either simple or with a little branching. The upright filaments may be arranged either at right angles to the prostrate filament (*Peyssonelia*) or arise at an acute angle and then curve to an upright orientation (*Cruoria, Petrocelis*). The upright filaments may either be held together relatively loosely by mucilage (*Cruoria*) or tightly compacted into dense pseudoparenchyma (*Peyssonelia*), with all intergrades occurring. All species of *Peyssonelia* appear to be calcified; in most this appears to be restricted to the horizontal filaments of the prostrate system, although in some species the thalli appear to be impregnated with calcium carbonate throughout (see p. 4). The thalli are attached to the substrate by short rhizoids which arise at intervals from the lower side of the prostrate filaments (Fig. 9, A, B). The variation in thallus aspect and structure is not very great in such genera (Denizot, 1968) and the principal criteria for taxonomic discrimination depend on the reproductive structures.

The coralline algae (Corallinaceae, Cryptonemiales) possess both purely prostrate crustose thalli and erect foliose thalli which arise from crustose bases often of some prominence. *Lithoporella* is probably the simplest of the crusts based upon the heterotrichous organisation. Species of this genus occur as epiphytes on various algae and marine phanerogams. In some, the vegetative thallus consists only of the prostrate filaments, erect filaments never being formed except in connection with the origin of reproductive structures. In other species, short erect unbranched filaments arise from the cells of the prostrate system. The presence of 'cap cells', characteristic of the Corallinaceae, is most easily demonstrated in species of *Melobesia* (Fig. 9, C, D, E). These cap cells are small cells which are situated in a terminal position on the upright filaments and in most cases are much smaller than the cells upon which they are located. The cap cells are formed prior to the development of an upright filament but, once formed, they never divide again. The growth of the upright filament is due to the meristematic activity of the *second* cell to arise from each cell of the prostrate system. Thus, the growth of the upright filaments is due to a form of intercalary cell division. The cap cells occur in all genera of the Corallinaceae and appear to have a function related to the deposition of calcium carbonate as well as serving to protect the meristematic cell, which is not strictly apical as in all other Florideophyceae, but sub-terminal.

Lithophyllum and *Lithothamnium* are widespread genera with species whose thalli, in certain cases, have an organisation similar to that of

Melobesia, although much more massive. In these species, the prostrate system consists of horizontal filaments which are closely compacted laterally. The cells of the prostrate system give rise to cap cells and then to cells which function, as in *Melobesia*, as the sites of meristematic activity, giving rise by their divisions to the upright filaments.

Both in the calcified Corallinaceae, as well as in the Peyssoneliaceae, there are some species and genera in which the crust is of a very different construction (Fig. 10). The thallus in these entities is not obviously of heterotrichous organisation, with a prostrate system of filaments attached to the substrate by rhizoids and with upright filaments produced on the upper surface. Filaments equivalent to the upright filaments are produced also on the lower surface in such genera as *Rhizophyllis* (Rhizophyllidaceae) and in some species of *Lithothamnium* (Corallinaceae), such as *L. lichenoides* and *L. muelleri*. It would appear completely incorrect to interpret such crusts on the basis of heterotrichous organization and an alternative interpretation is offered elsewhere (p. 83).

Hildenbrandia is a genus of uncertain position, sometimes referred to the Corallinaceae, or the Peyssoneliaceae, or assigned to a unique family of its own, with a massive encrusting thallus based on an obvious heterotrichous organisation. Most species are marine, although *H. rivularis* is widespread as a lithophyte in freshwater. The most commonly reported marine species, *H. prototypus*, has a thallus which persists for several years (Fig. 9, F). The upper layers are eroded, or grazed away by animals, from time to time and new growth is initiated from various sites on the surface, apparently at random, so that a section of the thallus shows various lenticular cell masses representing the bursts of growth which have occurred.

The erect thallus

The erect 'foliose' thalli of the Florideophyceae exhibit much greater diversity of form and organisation than do the encrusting thalli. By contrast, in the erect foliose thalli there may be differentiation at several levels of complexity, in addition to a primary heterotrichous organisation or in its absence (as in all members of the Ceramiales). The foliose thalli can be treated most conveniently by considering first those thalli in which a heterotrichous organisation is present, leaving the Ceramiales as a whole for later treatment. The occurrence of an initially heterotrichous organisation may often be difficult to establish in foliose thalli. In those instances where the prostrate system is well developed, with

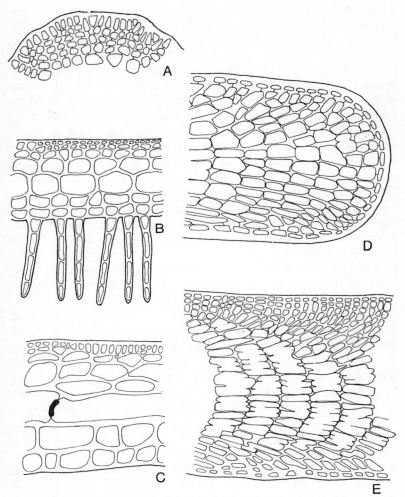

Fig. 10. *Crustose thalli where the structure is based on 'secondary hete-rotrichy' rather than a primary heterotrichous construction.* A–C, Rhizophyllis squamariae; A, *surface view of the thallus edge in the apical region* (× 480); B, *transverse section of mature thallus* (× 150); C, *longitudinal section* (× 320); D–E, Mesophyllum lichenoides; D, *section of the thallus edge in the apical region* (× 320); E, *section of mature thallus* (× 230)

(A–C *after Kylin;* D–E, *after Suneson*)

a number of upright filaments in addition to the one or more which form the basic core of the foliose thallus, there is usually no difficulty (Fig. 11). Problems of interpretation can arise where the prostrate system is relatively small. In such cases the breadth of the thallus may be of similar size to the diameter of the prostrate system, or even greater, so that the prostrate system forms a simple attachment disk at the base of the thallus. The heterotrichous nature of the latter can be demonstrated only by sectioning or by careful study of spore germination.

The simplest heterotrichous organisation is that found in many species of the *Acrochaetium/Rhodochorton* complex. A typical example is *Acrochaetium daviesii* (Fig. 14, A), in which the thallus is composed of a prostrate system of filaments from which arise the erect filaments. The prostrate system may be reduced in some species of this generic complex to the single cell formed by direct transformation of the spore from which the thallus has been derived. The erect filaments are branched in most species of the complex but usually in a somewhat irregular manner. The principal filament of the thallus and its lateral branches do not exhibit any indication of correlated growth and there is no differentiation between principal and lateral filaments. Growth of all filaments is by strict apical cell division.

A more elaborate type of construction is found in most foliose Florideophyceae, with a clear differentiation between principal and lateral filaments (Fig. 12). The latter often surround the former, the degree of aggregation varying from minimal (as in *Batrachospermum* (Fig. 14, B), *Dudresnaya* or *Nemalion*) to the more compact arrangements found in tough cartilaginous thalli. The principal and lateral filaments are differentiated not only physiologically but also morphologically, in many cases, with marked differences between the two types of filament. The apical cells may differ in terms of shape, size or mode of division, whilst their derivative products usually differ with respect to the degree of cell enlargement and thereby shape. It is often necessary to distinguish between the two types of filament in order to explain the patterns of cell division and cell enlargement by which a thallus develops. There are several different terminologies in use at the present time with reference to this feature, but none is particularly suitable. There has been considerable confusion in the application of terminology largely because it does not appear to have been appreciated that there are three levels of differentiation in Florideophycean thalli and the terminologies which have been used are not sufficiently discriminatory. Of the three levels of differentiation, the first is that in which there is no differentiation

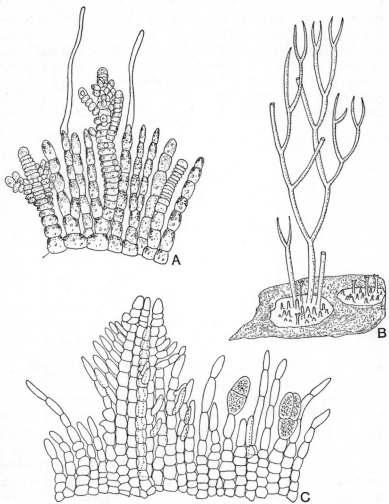

Fig. 11.   *Relationship of erect frond and crustose base.*   A, Gloiosiphonia capillaris, *origin of erect uniaxial fronds* (×300); B, Polyides
rotundus (×2); C, Platoma bairdii, *origin of erect multiaxial frond*
(×450)
    (A, C, *after Kuckuck;* B, *after Kützing*)

between filaments, as has been described previously for members of the *Acrochaetium/Rhodochorton* assemblage. The second level is that in which there is differentiation between a principal filament and surrounding lateral filaments. The aggregations of principal and lateral filaments are often of regular shape or they may undergo regular changes in shape as thalli age. It is convenient to refer to such an aggregation as an 'axis'. The third level of differentiation is that found in most complex thalli where there is not only differentiation between

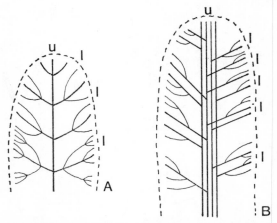

Fig. 12. *Differentiation of filaments within axes of erect Florideophycean fronds. A, uniaxial erect frond, with filament of unlimited growth (u) and filaments of limited growth (l); B, multiaxial erect frond, with filaments of unlimited growth (u) and filaments of limited growth (l)*
(A, B, *after Scagel*)

filaments in each axis, but also differentiation between axes. It is the occurrence of differentiation between axes (Fig. 13) which produces the highly regular, elaborately branched thalli and which contributes so much to their beauty and aesthetic appeal. For both axes and their constituent filaments, the basic feature of this differentiation involves control of growth. Obviously, the growth of an axis reflects the growth of the filaments of which it is composed. Nevertheless, it is necessary to have two systems of terminology referring to and discriminating between filaments and axes. In terms of behaviour, an axis is an integrated aggregation of filaments and represents more than their mere summation. It is convenient in discussions of growth control of filaments in an axis to refer to the principal central axial filament or filaments

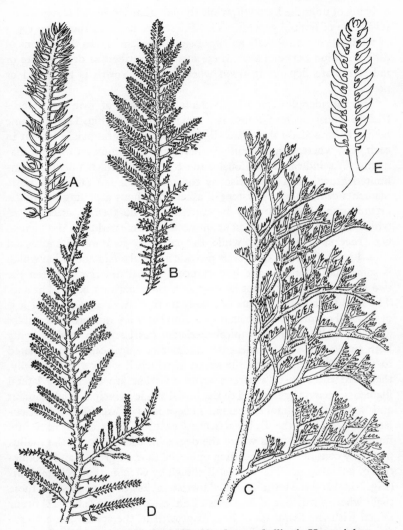

Fig. 13. *Patterns of axes in Florideophycean thalli.* A, Herposipho-
nia verticillata (×5); B, Laurencia splendens (×0.6); C, Microcladia
boreale (×1.6); D, Ptilota hypnoides (×1); E, Ptilota hypnoides (×4)
(A–E, *after Smith*)

as being of unlimited growth, while the surrounding lateral filaments are said to be of limited growth. The differences between these two types of filaments are essentially an expression of potential. In discussions of differentiation between axes, these can be described as determinate or indeterminate depending upon whether their growth is restricted or not.

The Florideophycean foliose thalli which are of primary heterotrichous construction fall into two major categories (Fig. 12). In some heterotrichous thalli there is a single principal central axial filament in each axis, whereas in other thalli there may be a number of such filaments. As indicated previously, these filaments are filaments of unlimited growth and they arise by the transformation of one or more filaments of the upright system of a heterotrichous germling where the organisation shows primary heterotrichy. The former situation, in which a single central axial filament occurs, is referred to as the unaxial construction (Fig. 12, A), while the latter is said to be multiaxial (Fig. 12, B). The two categories provide a useful characterisation which is sometimes of systematic importance. The distinction between the two types is frequently not as easy as various authors have indicated. The final form of the thallus is usually no indication as to whether it is of uniaxial or multiaxial construction, and there are numerous examples (Fritsch, 1945) where the apical meristem has been reported as being both single- or several-celled, the number of apical cells being related to the size of the axis and the speed at which it is growing. Finally, there are those cases where a series of what are essentially lateral filaments become aggregated to the initial single principal axial filament and 'webbed' by lateral adhesion. This process of 'webbing' is particularly apparent in the Delesseriaceae (see p. 78). It also occurs in a number of simpler cases where the degree of complexity of the thallus may vary considerably depending upon the rates of growth and production of lateral axes. The process is shown by *Gelidium* and *Rhodophyllis* (Fig. 15, F) and the term 'marginal meristem' has been applied to these thalli where a number of filaments of unlimited growth (i.e. axes) occur mixed with the filaments of limited growth in the apex. An extreme case of this phenomenon is shown in *Weeksia* (Norris, 1970) where the thallus has hitherto been regarded as being of multiaxial construction but where the multiaxial appearance follows from the aggregation of lateral filaments of unlimited growth and 'webbing'. One reason why the differences between uniaxial and multiaxial construction are not regarded in the present treatment with the significance accepted by many previous authors is that although the structural differences between the two are often considerable, the general principles of morphogenesis are the same in the two types of thallus (see p. 94).

There is one critical aspect related to the differences between multiaxial and uniaxial organisation which is frequently overlooked. The apical meristem of an axis of multiaxial construction contains a number of apical cells, usually eight or more. This apical meristem can divide into two parts, which may be equal or unequal, thereby undergoing a truly dichotomous branching. By comparison, the apical meristem of an axis of uniaxial construction contains a single apical cell. The mode of division of apical cells in Florideophycean thalli is such that a single apical cell *cannot* separate into two parts, as occurs in the brown alga *Dictyota*. Even though thalli of uniaxial construction frequently have a dichotomous aspect, as in most species of *Ceramium*, for instance, the pattern of branching of a uniaxial axis *cannot* be dichotomous. The sequence of developmental stages entails the formation of a *lateral* axis and then the rotation of both the lateral and the parent axis to give the dichotomous appearance. This rotation is termed 'evection'. It is not made clear in most texts that the mode of branching in a thallus of dichotomous aspect need not be and often is not true dichotomy. This poses many problems as 'dichotomous' and 'not dichotomous' are favourite criteria for separation in keys and it should be appreciated that these terms are usually applied to appearance rather than precisely to the mode of branching.

Acrochaetium daviesii, discussed previously, is a thallus of uniaxial construction (Fig. 14, A), while *Batrachospermum* is a more complex example of the same type, with the thallus differentiated into filaments of limited and unlimited growth (Fig. 14, B). The apical cell of the filament of unlimited growth is thimble-shaped and it divides transversely. Each segment formed by such division gives rise to a number of primordia, usually four, which function as the apical cells of filaments of limited growth. Once formed by the division of these apical cells, the filaments of limited growth form a loose whorled aggregation, radiating from the single central axial filament of unlimited growth. In addition to the radiating filaments of limited growth, some of the latter exhibit a different growth habit, in that they grow down either adpressed or in close proximity to the axial filament (Fig. 14, C, D). The degree of investment and coverage depends on the number and extent of growth of the two types of filaments of limited growth. Where the radiating filaments of limited growth in a thallus are only loosely

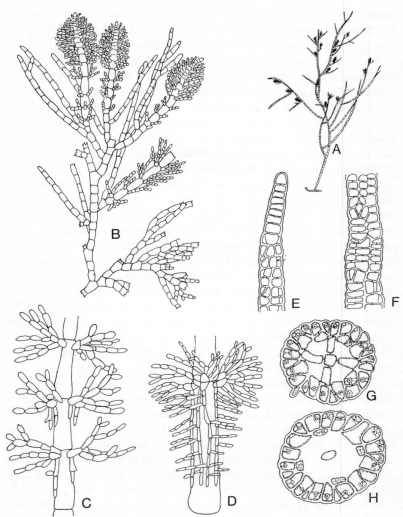

Fig. 14.   *Simpler examples of uniaxial organisation.*   A, Acrochaetium daviesii, *whole plant* (×27); B, Batrachospermum ectocarpum,
with origin of erect fronds from Chantransia *stage* (×150); C, Sirodotia
suecica, *young axis showing radiating and downgrowing filaments of
limited growth* (×300); D, Sirodotia suecica, *older portion* (×150); E,
Lemanea fluviatilis, *apex of upright frond* (×360); F, Lemanea
fluviatilis, *later stage* (×360); G, Lemanea fluviatilis, *transverse section
of young frond showing the pericentral cells connecting the axial cell to
tubular portion of the thallus* (×360); H, Lemanea fluviatilis, *section
of frond other than at position of pericentral cells* (×360)
   (A, *after Taylor;* B, *after Sirodot;* C–H, *after Kylin*)

aggregated the downgrowing filaments may be readily visible and any changes in the balance between the two types of filaments of limited growth can have a considerable effect on the form of the thallus (Fig. 14, C, D). The shape and size of the cells in the downgrowing filaments of limited growth may be very different from those in the radiating filaments. That the radiating and downgrowing filaments of limited growth are equivalent to one another is shown by the frequent interconversions between the two types of filaments of limited growth. The downgrowing filaments are often referred to as 'corticating filaments', a term better avoided. The downgrowing filaments may not be particularly easy to detect in more compact thalli where they are often composed of very elongate cells. The terms 'hyphae' or 'rhizines' are frequently applied to the latter but there is no justification for any nomenclatural distinction between these and segmented downgrowing filaments of limited growth as there is no strict morphological delimitation between the two types. The basic organisation found in *Batrachospermum*, in which a segment formed by the division of an apical cell produces four lateral primordia which serve as the apical cells for filaments of limited growth, occurs in a large number of genera. Some of these are as soft as the loosely aggregated filaments comprising the thallus of *Batrachospermum*, although many examples of this basic type of organisation are extremely compact. The thallus of *Batrachospermum* is essentially radial in symmetry, the filaments of limited growth arising from the four primordia being of equivalent development. In many genera where the thallus is more compact, there is frequently an isobilateral symmetry. The first two primordia are cut off on the lateral flanks of the segment formed by the division of the apical cell and these then give rise to filaments of limited growth of some length. The second pair of primordia produces very much shorter filaments which contribute essentially to the depth of the thallus. The filaments thus give a thallus which is of greater breadth than depth. Growth of the filaments of limited growth is by the division of the apical cells of those filaments and the subsequent enlargement of the products. The segments produced may divide to produce laterals, but they will never divide again in the same plane as that by which they were formed. In such compact thalli, the apical cells aggregate to form the superficial layer of cells over the thallus. Thus, the boundary layer of cells over the whole surface of the thallus is not an inert 'epidermis', as is so often assumed, but a region of apical cells, which may be dividing actively or, if not, then at least potentially capable of resuming meristematic activity. This aspect of meristematic organisation and the ease with which meristematic activity can be resumed, common to all Florideophyceae, is probably the cause of the frequency with which proliferations can arise in such algae, the ready effects of environmental conditions on the shape of the thallus caused by changes in the division rates of the apical cells of the filaments of limited growth, and the marked effects of animal grazing and fungal attack.   The *Batrachospermum* type of thallus organisation can also give rise to tubular thalli, as in *Lemanea* or *Dumontia* (Fig. 14, E-H).   The filaments of limited growth arise from four primordia, as in *Batrachospermum*.   The lowermost cells of these filaments of limited growth elongate in a radial direction, while the remaining cells aggregate.   A continuous tube of cells is thus formed, traversed by the central axial filament, with each cell of the latter attached to the tube by the elongate lowermost cells.   The organisation of *Dumontia* and *Gloiosiphonia* is essentially similar to that of *Lemanea*.

A slightly more elaborate pattern of apical segmentation occurs in thalli of several genera, although these are similar in many respects to the basic organisation of *Batrachospermum*.   In these more elaborate thalli, the apical cell of the single filament of unlimited growth divides obliquely, rather than transversely as in all the examples which have been considered previously.   The segments formed by such oblique divisions are wedge-shaped and each segment gives rise only to a single primordium of a filament of limited growth, usually on its outer, larger face.   Obviously the orientation of the succesive filaments of limited growth is determined by the obliquity of the apical cell divisions. This may be simply alternate as in *Cryptosiphonia* or *Cystoclonium* (Fig. 15, D), although in *Plocamium* the apical cells of a major axis may segment in three planes with divisions only in two planes in some of the smaller axes.   The orientation and obliquity of the apical cell divisions may exhibit spiral arrangements of some complexity, as in *Naccaria* and *Bonnemaisonia*.

Turning next to a consideration of Florideophycean thalli in which the axes have a multiaxial organisation (Fig. 16), the thallus of *Nemalion* represents the classic example of this type of construction.   The filaments of which the thallus is composed are loosely aggregated and embedded in mucilage, so that examination is particularly easy.   Each axis contains up to 250 filaments of unlimited growth which form a compact central mass from which the filaments of limited growth radiate. The first-formed cells of the filaments of limited growth enlarge considerably.   The last-formed, lying immediately behind the apical cells

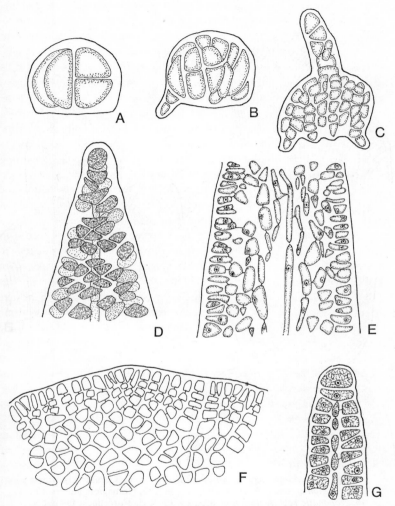

Fig. 15. *More complex examples of uniaxial construction.* A–E,
Cystoclonium purpureum; A–C, *stages in thallus development from a*
spore (× 400); D, *apex of mature plant* (× 475); E, *cross section of mature*
plant (× 350); F–G, Rhodophyllis divaricata; F, *surface view of the edge*
of the apical region (× 600); G, *section of apical region* (× 600)
   (A–G, *after Kylin*)

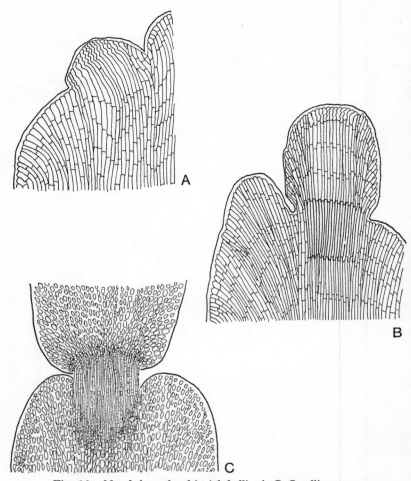

Fig. 16. *Morphology of multiaxial thalli.* A–C, Corallina *sp. stages in formation of joints* (A, B, ×900; C, ×60); D, Cumagloia andersonii, *diagrammatic representation of apex* (×150); E, Furcellaria fastigiata, *longitudinal section of apex* (×100); F, Chylocladia verticillata, *longitudinal section of apex* (×200); G, Callophyllis edentata, *transverse section of thallus* (×70); H, Furcellaria fastigiata, *transverse section of thallus* (×55); I, Rhodymenia palmata, *transverse section of thallus* (×125)

(A–C, E, H, *after Oltmanns;* D, *after Smith;* F, G, *after Kylin;* I, *after Taylor*)

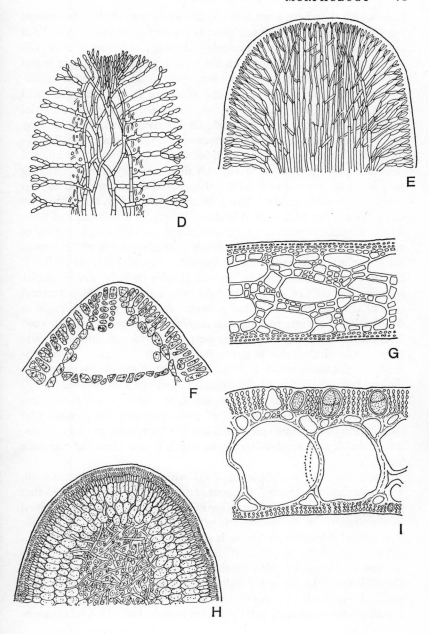

D

E

F

G

H

I

from which they have been formed, do not enlarge to the same extent but form a compact bushy cell mass at the surface of the thallus. Each bushy cell mass is composed of relatively short cells and adjacent cell masses pack together to form an almost complete surface layer to the thallus. *Nemalion* shows radial symmetry, as do many other multi-axial Florideophyceae in which the structure is more compact and cartilaginous (*Agardhiella, Ahnfeltia, Polyides*). Various multiaxial thalli are either compressed or markedly flattened, particularly in many of the genera referred to the Cryptonemiales, Gigartinales and Rhodymeniales (*Chondrus, Gigartina, Kallymenia, Rhodymenia*). The internal structure and organisation of all such thalli are very similar, differing only in minor matters such as cell shape, size, degree of secondary pit connection formation, etc. These differences, however, are sufficiently characteristic for sectioning to be the only way by which many of these foliose Florideophyceae may be recognised. The central portion of such thalli is composed of elongate cells between some of which abundant secondary pit connections are formed. Superficially, the thallus is bounded by the filaments of limited growth and the perpendicular filaments are cells of intermediate size which represent the lowermost cells of the filaments of limited growth. Attempts are often made to differentiate thalli into regions such as medulla and cortex. While such discrimination may be helpful for purposes of identification it should be appreciated that this is not a fundamental difference. Developmentally, the tissue which is in the inner part of the thallus represents the product of division of the apical cells of the filaments of limited growth at an early stage of development, the tissue which is now immediately behind the apical cells of those filaments was produced by the division of the same apical cells but at a later point in time. Thus, it is probable that much of the tissue which is now 'medulla' was once 'cortex'.

There are various multiaxial thalli which are markedly calcified. The organisation of both *Galaxaura* and *Liagora* is very similar to that of *Nemalion*, except that the surface of the thallus is markedly calcified in the vicinity of the apical cells of the filaments of limited growth. The central axial core of *Liagora* is also said to be irregularly calcified in older parts of the thallus. A multiaxial structure associated with marked calcification is characteristic of the erect foliose thalli of the Corallinaceae. In all such coralline algae, the thalli are composed of calcified segments separated by flexible non-calcified joints. In *Corallina* and other such thalli, the axes contain 100–250 axial filaments which are composed of elongate cells (Fig. 16, A-C). The filaments of limited growth arise from these and are clearly demarcated from them. The growth of the filaments of limited growth is by a specialised form of cell division similar to that described previously in some of the calcified encrusting thalli, the outermost cell being a small, non-dividing 'cap cell'. At a joint, the thallus consists only of the axial filaments of unlimited growth, the cells of which are particularly elongate in that region and devoid of any associated filaments of limited growth. The deposition of calcium carbonate takes place particularly around the cap cells and between the filaments of limited growth so that the joint remains devoid of calcification. The axes are cylindrical in some genera of the Corallinaceae (*Corallina, Jania*) or markedly compressed in others (*Calliarthron, Bossiella*).

Two types of tubular organisation are found in Florideophyceae with multiaxial thalli. The first, which occurs in *Scinaia, Pseudogloiophloea* and other genera of the Chaetangiaceae, is analogous to that described previously for the uniaxial genera, *Lemanea* and *Dumontia*. The mass of axial filaments in *Scinaia* forms a central core which becomes separated from the compact outermost tissue formed by the aggregation of the terminal (i.e. apical) portions of the filaments of limited growth as a result of the extreme elongation of the lowermost cells of the latter. A second type of tubular thallus based upon a multiaxial organisation occurs in certain genera of the Rhodymeniales. In *Lomentaria* and *Chylocladia*, the filaments of unlimited growth in the tubular portions of the thallus do not form a solid core (Fig. 16, F), as in *Nemalion* and the other examples which have been considered previously, but are themselves arranged as a hollow tube. The filaments of limited growth are very short and lie outside the tube formed by the filaments of unlimited growth. The hollow tube in the centre of the thallus is completely unconstricted in certain species (*Lomentaria clavellosa*) whereas in others it is constricted at intervals. The constrictions in *Chylocladia verticillata* and *Champia parvula* are associated with plates of tissue, the so-called 'diaphragms', which are formed at intervals across the cavity of the tube. These diaphragms develop from primordia which arise on the inner sides of the filaments of unlimited growth, growing across the cavity of the tube and connecting together the filaments of unlimited growth. The thalli of *Lomentaria articulata* are also constricted at intervals but these constrictions do not represent the positions of diaphragms, but rather solid areas where the filaments of unlimited growth adhere laterally rather than form a tube.

The main axes and basal parts of the branches of *Gastroclonium ovatum* and *G. coulteri* remain solid throughout, the tubular arrangement being restricted to the terminal portions. The organisation found in *Lomentaria articulata* and *Gastroclonium* probably represents an indication of the way by which the hollow, tubular thallus of the Champiaceae arose from the more usual pattern of multiaxial organisation found in other Florideophyceae.

All the morphological examples which have been considered so far exhibit a greater or lesser degree of primary heterotrichy. The remaining genera to be considered, all of which are referable to the Ceramiales, are characterised by a total absence of any indication of primary heterotrichy. The order Ceramiales is exclusively uniaxial, all spores on germination display a bipolar orientation from the earliest stages of development and primary heterotrichy is totally lacking (Fig. 17). The thalli of species of *Callithamnion* (Fig. 17, A) are probably the most simple from the morphological point of view. The thalli are much branched, the principal axes and lateral branches are all uniseriate, uniaxial structures, the lateral branches arising (with rare exceptions) alternately from every cell and arranged either distichously or in spiral arrangements of varying degrees of complexity (Rosenvinge, 1920). There is no sharp morphological differentiation between filaments of limited and unlimited growth although there are obviously physiological differences. There is between the different orders of lateral branches a distinct hierarchy of dominance. Downgrowing filaments, which are modified filaments of limited growth, are present in various species. These arise from the lowermost cells of the lateral branches as well as from the cells of the principal axes. The extent to which these downgrowing filaments develop varies considerably. In some species they are represented by a few short single cells, whereas the most luxurious development occurs in *C. tetricum* where it may achieve a depth of from six to eight cell layers. *Antithamnion* (Fig. 17, D) is similar to *Callithamnion* in many respects, although the branching is opposite as opposed to alternate. The lateral branches arise in *Antithamnion* in whorls of two, three or four and the demarcation between filaments of limited and unlimited growth is much greater than in *Callithamnion* (Fig. 17, A, D). *Ptilota* (Fig. 17, B) is an excellent example with which to consider the different interactions between filaments and axes which have been mentioned previously on a number of occasions (pp. 60, 63). Every cell of the principal axes gives rise to four primordia. Of these, each primordium of an opposite pair on *every other cell* of a principal axis forms a lateral filament of unlimited growth; the remaining primordia give rise only to filaments of limited growth. The latter aggregate to give a compact mass of cells although some function as downgrowing filaments as in *Ceramium* or *Batrachospermum*, etc. The pattern of segmentation in the lateral filaments of unlimited growth is essentially similar to that of the principal axes on which they are formed. The number of orders of branching depends upon the luxuriance of the specimen. The two lateral branches of unlimited growth formed by the two opposite primordia of a pair develop in a most asymmetric manner, so that one is several times longer than the other when growth ceases. In addition, the long and short lateral axes are arranged alternately along the principal axis but these are of equivalent development for considerable distances so that the outline of the frond is highly regular. Most of the lateral axes are arranged in this highly determined manner and in undamaged thalli only a very few are indeterminate, showing no restriction of growth and repeating the organisation of the primary axis in every respect. Damage, as a result of wave action or animal grazing, will cause the hitherto determinate lateral axes to recommence growth and the regular outline of the frond is largely lost in consequence. A similar series of interactions to those observed in *Ptilota* can also be demonstrated in the thalli of various species of *Ceramium*. The pattern of development in these is similar to that described for *Ptilota*, although differing in certain specific points of detail. Each segment formed by the division of the apical cell of a principal axis in *Ceramium* (Fig. 17, F) gives rise to a number of primordia. The number of primordia varies from species to species from a minimum of four to a maximum of twelve. Even in those cases where only four primordia are produced from a segment, these are not formed in two opposite pairs as in *Ptilota*, but by a regular sequential development. The position of the first primordium is determined by the preceding 'dichotomy' of the axis, in that the first pericentral cell is always on the outer face of the axis or lateral, relative to the preceding 'dichotomy', and the other primordia are then formed in peripheral sequence in pairs. Each primordium gives rise to filaments of limited growth, usually four in number, of which two grow upwards and two grow downwards. The upward- and downward-growing filaments of limited growth aggregate to form a compact ring of tissue. Adjacent rings of tissue may remain discrete, occurring at intervals along the uniseriate axial filaments or they may meet or even fuse to form a complete investment to the axial filament. Although the thallus of *Ceramium*

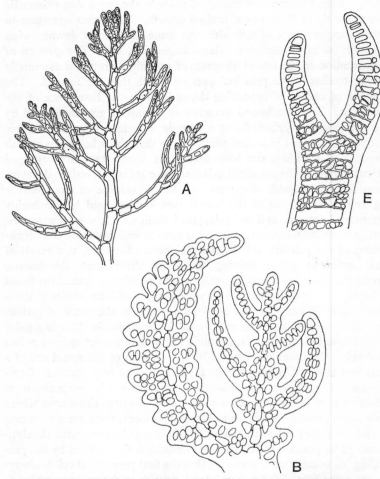

Fig. 17. *Morphology of Ceramiales—Ceramiaceae.* A, Callithamnion hookeri (×70); B, Ptilota serrata (×320); C, Plumaria elegans
(×320); D, Antithamnion plumula, *with positions of secretory cells
indicated* (×200); E, Ceramium *sp.,* *showing development of separated
'cortical bands'* (×200); F, Ceramium rubrum, *showing adhesion of
adjacent 'cortical bands'* (×250)

   (A, *after Rosenvinge;* B, C, E, F, *after Taylor;* D, *after Feldmann-
Mazoyer*)

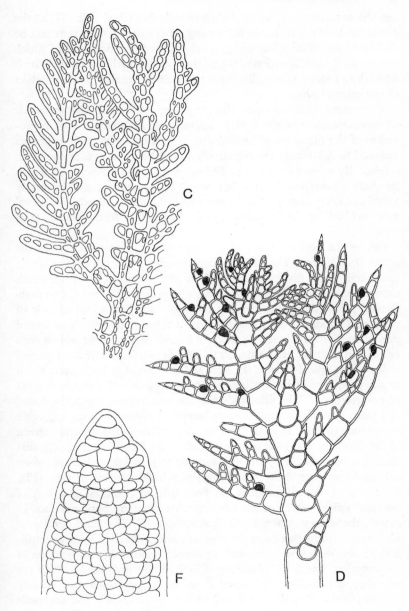

has the appearance of being dichotomously branched (Fig. 17, E) the actual mode of branching in this or any other uniaxial thallus cannot be dichotomous. The apical cell gives rise to two products, one of which represents the continuation of the original axis, the other a lateral branch which by a process of evection turns to resemble closely the continuation of the original axis.

The genera *Callithamnion*, *Antithamnion*, *Ptilota* and *Ceramium* are all representatives of the family Ceramiaceae. Although some aggregation of the filaments of limited growth can occur in thalli of genera assigned to this family the organisation is still obviously filamentous in origin. By comparison, in the Delesseriaceae, the filaments of which the thalli are composed adhere laterally. This structure resembles closely a thallus such as that of *Antithamnion* with the principal axis and lateral axes 'webbed' by the filaments of limited growth to form a flat sheet of tissue.

One of the best examples of this type of organisation occurs in the genus *Hypoglossum* (Fig. 18, A). Each principal axis possesses a single apical cell, which divides transversely to give segments which are much shorter than broad. These then give rise to four primordia, two first-formed primordia which develop laterally and then a second pair of primordia produced on the dorsal and ventral surfaces. The lateral primordia divide obliquely to produce laterals of first order and on each cell of this filament are borne, on the abaxial surface, laterals of second order. The apical cells of the first and all second order laterals reach to the periphery of the thallus thus indicating that their growth must be co-ordinated. The principal axial filament, together with the dorsal and ventral primordia and the products of division of these, aggregate to produce the midrib of the thallus. In *Delesseria* the general pattern of development is similar to that described for *Hypoglossum* except that the segments formed by the division of the principal apical cell of an axis undergo intercalary cell division of a highly organised type (Fig. 18, B, C). A segment once it has been formed functions as a sort of 'internal apical cell', cutting off segments from its basipetal face in exactly the same way as the principal apical cell. The thalli of *Crypto-pleura* are even further from the regular pattern of cell division exemplified by *Antithamnion*. Although obviously uniaxial in construction in the sporeling stage, the filaments of limited and unlimited growth are not morphologically differentiated as in *Hypoglossum* but merely exhibit physiological differentiation as in *Callithamnion*. The net result is to produce a thallus which may be interpreted as a 'webbed' structure,

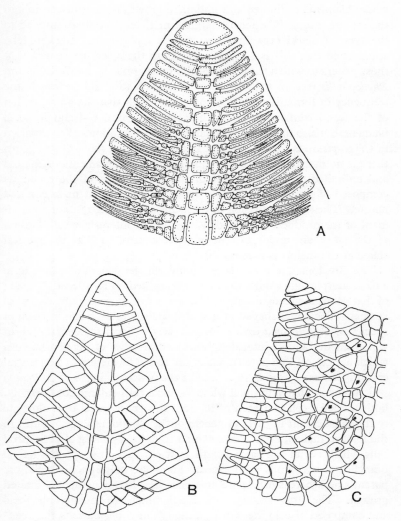

Fig. 18. *Morphology of Ceramiales—Delesseriaceae.* A, Hypo-glossum woodwardii, *apex* (×600); B, Delesseria sanguinea, *apex* (×560); C, Delesseria sanguinea, *later development indicating positions* (★) *of intercalary cell divisions* (×425)
(A, *after Kylin;* B, C, *after Rosenvinge*)

where filaments of limited and unlimited growth are not discernible in the mature stage. The term 'marginal meristem' has been applied to the type of apical organisation seen in *Cryptopleura*. It should be appreciated that this is very different from the apical construction of those genera to which the term has also been applied. In *Gelidium* and *Rhodophyllis* (see p. 69), the term is applied to thalli in which the frequency of formation of lateral axes is so high that the apical region of the frond contains not only the principal apical cell and its axial filament, but also those of several lateral axes in addition. The situation in *Cryptopleura* is totally different in that in mature thalli of this genus there is no morphological differentiation between filaments of limited and unlimited growth. Because of this, the margin of the thallus is composed of the apical cells of a series of filaments of undetermined nature. The rates of division of these apical cells, the rates of enlargement of the products of their division and the intercalary cell division of the latter are under extremely strict control so that the general shape of the thallus is maintained.

The Rhodomelaceae is a large family, all the genera of which have a basic morphology similar to that of *Polysiphonia* (Fig. 19, A-C). Thalli of that genus are recognisable by the characteristic arrangement of cells in tiers. The dome-shaped principal apical cell of each axis in thalli of species of *Polysiphonia* divides transversely to form segments, each of which divides almost immediately to form a series of pericentral cells which are cut off in a pattern analogous to the formation of pericentral cells in *Ceramium* (see p. 76). Unlike the pericentral cells of *Ceramium* which remain relatively small while the axial cells enlarge considerably, both the axial cell and each of its derivative pericentral cells elongate to the same extent in thalli of genera of the Rhodomelaceae. Thus, in the mature thallus, each axial cell is surrounded by a ring of pericentral cells which are of the same length as itself. In most species, lateral branches of two kinds are formed. Some are polysiphonous like the parent axis, whereas others are uniseriate hair-like structures of limited growth. The latter are the trichoblasts which are usually ephemeral and composed of colourless cells, although in *Brongniartella* they contain chloroplasts and are relatively long-lived. The trichoblasts usually bear the gametangia and in some species of *Polysiphonia* only gametangial-bearing trichoblasts are formed although this is not the case with all species. In some species of *Polysiphonia*, downgrowing filaments develop from the basal poles of the pericentral cells and invest the axis in a heavy filamentous covering. These down-growing

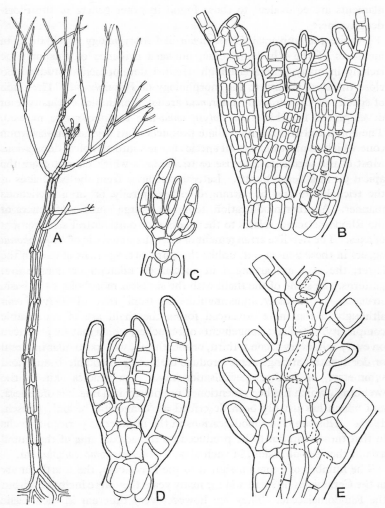

Fig. 19. *Morphology of Ceramiales—Dasyaceae and Rhodomelaceae.*
A, Polysiphonia violacea, *young plant showing trichoblasts and scar cells
remaining after they have been shed* (× 130); B, Polysiphonia nigrescens,
apex (× 400); C, Polysiphonia nigrescens, *trichoblasts* (× 400); D,
Rhodoptilum plumosa, *apices showing sympodial development* (× 600);
E, Rhodoptilum plumosa, *more mature axis with no evidence of sympo-
dial origin* (× 400)
  (A, *after Rosenvinge;* D–E, *after Rosenberg*)

filaments are equivalent to those found in other genera of the Florideophyceae.

Despite the universal *Polysiphonia*-like morphology of all thalli in
members of the Rhodomelaceae, outward evidence of this may be
lacking in many genera although detailed developmental studies dis-
close the clear affinity to the morphology of *Polysiphonia*. The thalli
of species of *Chondria* and *Laurencia* are good examples of this type of
development in which the *Polysiphonia*-like morphology is masked.
The apical cell and the immediate products of its segmentation form a
cone which in most but not all species is sunk into an apical depression.
Most of the segments give rise to trichoblasts which emerge from the
apical pit in a dense bunch.   Lateral axes arise from the basal cells of
the trichoblasts and are arranged either spirally or in a distichous
manner.   Another modification, found in a large number of genera of
the Rhodomelaceae, relates to the bilateral or dorsiventral arrangement
of axes.   The tier-like arrangement of cells characteristic of *Polysiphonia*
occurs in these genera but, unlike the spiral arrangement of axes in the
latter, the axes are arranged in specialised bilateral or dorsiventral
patterns.   In some of the thalli with the simplest morphology the basic
structure is virtually indistinguishable from that of *Polysiphonia*
although in the more advanced forms the thalli are of remarkable
complexity. · Special arrangements of branches, which may be produced
on every cell, every second, third, or fourth cell, with particular bilateral
or dorsiventral arrangements produce thalli of remarkable beauty and
symmetry.   A further specialisation involves a process akin to the
'webbing', which has been mentioned previously in a number of genera,
and which has usually been described as 'congenital fusion' (Fritsch,
1945).   This can cause modifications in the number of pericentral cells
in that these tend not to be produced on the adaxial side of the lateral
axis in the area of fusion, in such algae as *Pterosiphonia complanata*.

The number of genera assigned to the Dasyaceae, the fourth family
of the Ceramiales, is small and for many years they were included within
the Rhodomelaceae.   They are however very different in their basic
pattern of development, in that this is based on a sympodium in genera
of the Dasyaceae (Fig. 19, D, E).   The main axis of the thallus is made
up of the basal portions of successive laterals.   In some genera, such
as *Daysa* or *Rhodoptilum*, each lateral contributes only a single cell to
the principal axis because the growth of that main axis is continued by
a lateral formed on the lowermost cell.   In other genera, the continuation of the principal axis is by a lateral formed on the second or

third cell.   The principal axis becomes polysiphonous while in *Hetero-siphonia* the lowermost segments, usually the lowest three, also develop in this way, the apical portion of each lateral filament remaining in a uniseriate condition.

The occurrence of 'secondary heterotrichy'

One critical feature which appears to have been almost completely overlooked by previous investigators is that whether or not the thallus of a Florideophycean alga shows a primary heterotrichous arrangement of filaments at the base, the thallus as a whole may be differentiated secondarily into axes which are erect and those which are prostrate.   A convenient term for this phenomenon is 'secondary heterotrichy of axes' although it should be appreciated that there are certain objections to this phrase.   This phenomenon occurs in diverse thalli of genera of the Florideophyceae.   In certain cases, such as *Cryptopleura ramosa* or *Plocamium cartilagineum*, thalli may be either erect or prostrate and the mode of growth appears to be determined by external conditions. Contact stimulates the development of attachment structures (see p. 39) along the length of the thallus and the degree of illumination also appears to be critical.   Such 'secondary heterotrichy' appears to be highly significant in the phenomena of perennation (see p. 211).   On the other hand, there are certain species which appear to have adopted permanently a status equivalent to 'secondary heterotrichy'.   The structure of the species of *Lithothamnium* mentioned previously (p. 58) differs markedly from the usual primary heterotrichous arrangement of filaments found in encrusting thalli.   Such a thallus does resemble closely an axis of, say, *Amphiroa* laid upon its side.   A similar relation-ship may be detected between thalli of *Rhizophyllis squamariae* (Fig. 10, A–C) and *Rhodophyllis*.   It has been argued (Kylin, 1937a) that the prostrate habit in *Rhizophyllis, Coriophyllum* and *Ethelia* is 'secondary' whereas the present view extends this argument.   These hypotheses could easily be tested by following in detail the development of such a thallus from a spore but this does not appear ever to have been under-taken.

Parasitism and epiphytism

Red algae occur on and in a variety of living and non-living substrates. Many occur as epiphytes which are attached only superficially to the host species on which they develop.   There is evidence (Linskens, 1963a) that even with such a superficial relationship, P^{32} may move from host

D

to epiphyte and, to a lesser extent, from epiphyte to host. The balance between movement in the two directions varies from example to example. Certain host species are characterised by an abundance of epiphytes while other species are almost always devoid of them. It has been suggested that these differences in the degree of epiphytic development may be a reflection of the surface tension of the host (Linskens, 1963b). Measured values for surface tension in red algae range from 62·4 ergs/cm² in *Phyllophora nervosa* to 355·2 ergs/cm² in *Sebdenia monardiana*. It does not appear to have been investigated whether there is a chemical basis for such measurements in addition to, or as a replacement of, this physical feature. It would be a most attractive hypothesis to suggest that the host species was providing either some stimulation for initiating spore germination or some material necessary for further development but there is no evidence to support these ideas.

In some cases, species have been reported occurring as epiphytes or endophytes only on a single host species. Although host specificity has been used as a criterion for taxonomic discrimination, particularly in the *Acrochaetium/Rhodochorton* assemblage, this feature must be regarded with caution. Reports of host specificity are often based on a very restricted number of collections (Woelkerling, 1971). Experimentally (White and Boney, 1969; Woelkerling, 1971), some of the endophytes have been grown *in vitro* away from the host species, for periods in excess of one year. *Acrochaetium asparagopsis* and *A. infestans*, which are reported from red algal hosts other than *Heterosiphonia plumosa*, germinated on the surface of that species but failed to penetrate and merely produced superficial epiphytic growths, while spores of *Acrochaetium endophyticum* obtained from *Heterosiphonia* were able to produce endozoic filamentous growths within a hydroid organism. Such studies indicate the need for detailed investigations of host specificity and it is clear that until more information is available its use as a taxonomic criterion is open to question. There is in the red algae a complete gradient of forms whose relationship ranges through obligate epiphytism, in which an epiphyte occurs only on one or more specific substrate species, to those interactions which are assumed to be of a more specialised nature in which there is a deep penetration of the host by endophytic tissue and a reduction in the superficial structure or photosynthetic pigmentation of the epiphyte/parasite.

One species which probably falls into the intermediate category is *Polysiphonia lanosa*, which occurs on *Ascophyllum* and other Fucaceae (Fig. 20, A). The initial primary rhizoid is a bulbous structure which

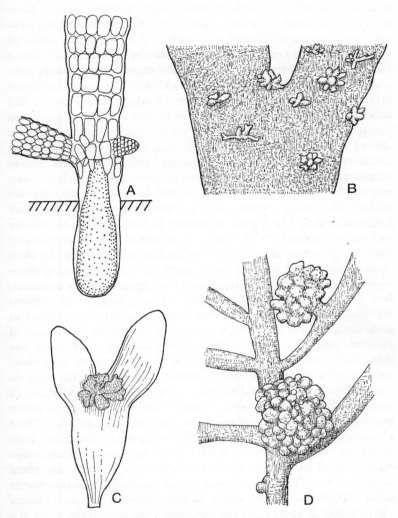

Fig. 20. *Parasites and obligate epiphytes.* A, Polysiphonia lanosa, *on* Ascophyllum nodosum (× 150); B, Faucheocolax attenuata, *on* Fauchea laciniata (× 2.5); C, Gonimophyllum buffhamii, *on* Crypto-pleura ramosa (× 5); D, Gardneriella tubifera, *on* Agardhiella coulteri (× 4)

(A, *after Sauvageau;* B, D, *after Smith;* C, *after Newton*) can enter the substrate species either through the cuticle, sites of damage or the meristematic apices. There have been various claims as to the nature of the relationship between *P. lanosa* and the substrate species and it has been claimed on the basis of histochemical staining that the bulbous rhizoid acts in a haustorial capacity, withdrawing material from the substrate species (Batten, 1923). The tip of the rhizoid may develop a protoplasmic protrusion (Rawlence and Taylor, 1970) although there is no evidence that this formed any sort of haustorial contact with the substrate. It was shown that the original primary rhizoid may become bulky through the addition of secondary rhizoidal extensions from adjacent cells. Secondary rhizoids may also arise at some distance from the primary rhizoid and these may either be bulbous or resemble those of other species serving to attach to the substrate surface rather than penetrate. The question of the status of this species has been confused by various statements that it can occur directly on rock and not as an epiphyte. Further investigation (Dixon, unpublished data) indicates that there are various explanations for this. In some cases specimens are growing epiphytically on the basal crust of *Ascophyllum* which can become fragmented with age and, in this condition, is indistinguishable from the underlying rock without very careful examination. The capacity to form 'normal' attachment rhizoids at some distance from the primary bulbous rhizoid means that a plant of *P. lanosa* may have been attached to the Fucaceous substrate at one point but careless collecting can leave this and remove only those portions with 'normal' rhizoids. Similar bulbous rhizoids occur in other epiphytic species of *Polysiphonia*, such as *P. rubrorhiza* and *P. saccorhiza* (Hollenberg, 1968), although there is no evidence as to the physiological relationship of these with the substrate species. Those Florideophyceae which are more deeply penetrating into the substrate species and which exhibit reductions of morphology or pigmentation provide better indications of a relationship which is usually described as parasitism. *Ceratocolax hartzii* occurs on *Phyllophora truncata* as small pinkish tufts. The degree of entry into the substrate species varies in that in some cases the *Ceratocolax* merely spreads below the cuticle whereas in other instances the degree of penetration is deeper and filaments spread extensively through the medulla of *P. truncata* causing a certain amount of tissue breakdown. The thalli of various species of *Gonimophyllum*, which occur on *Cryptopleura*, *Botryoglossum* and other members of the Delesseriaceae, possess several features resembling those of *Ceratocolax* (Fig. 20, c). Pale pink fronds (to 1 cm) arise in tufts from the tissue of the substrate species which undergoes a certain degree of hypertrophy. Both in *Ceratocolax* and *Gonimophyllum* a certain amount of photosynthetic pigmentation is produced although the quantity is much less than normal for a red alga, the thalli are relatively minute and there is extensive development of penetrating endophytic filaments. These observations are used as arguments to justify a parasitic relationship between the two organisms. Photosynthetic pigments appear to be lacking in *Gardneriella*, *Holmsella* and *Harveyella*. *Gardneriella tubifera* occurs on *Agardhiella coulteri* (Fig. 20, D) whereas *Harveyella mirabilis* occurs on *Rhodomela confervoides* and *Holmsella pachyderma* occurs on *Gracilaria verrucosa*. In all cases, endophytic filaments penetrate the tissue of the substrate species and secondary pit connection formation occurs between the substrate species and the supposed 'parasite'. The greatest degree of morphological reduction occurs in *Colacopsis* and *Choreonema* which are completely colourless and where the entire plant is endophytic except for the reproduction stages which occur above the surface of the substrate species.

At the present time, although the assumption that there is some degree of parasitism occurring in these Florideophyceae is generally accepted, the primary evidence in favour of such an interpretation is extremely weak. There is a desperate need for a complete investigation, both at the biochemical and fine-structural levels, of the whole range of forms from the indiscriminate epiphytic species of red algae to those where the morphology and pigmentation are much reduced and where a more specific relationship occurs with the substrate species. It is most interesting to note that the examples of the latter are almost all associated with species of relatively close affinity, usually of the same family. Such a close taxonomic affinity suggests that examples of this relationship might well provide useful material for studies of the DNA and enzyme relations of host and associated species, whether or not 'parasite' is the correct term to be applied. It is also possible that the apparently close affinity between host species and 'parasite' may mean that the 'parasite' is in fact nothing more than a gall (see p. 42).

Embryology

The adult thallus of a red alga, such as has been described previously, is rarely formed directly from a spore and there are usually intermediate stages in development. Some account of these intermediate developmental stages is necessary and this is best presented in the form of a comparative survey of the red algae as a whole rather than as a series

of disconnected examples.   The use of the term 'embryology' for the
description of early developmental stages in a group of algae demands
some explanation.   The original botanical usage (1728) of the term
'embryo', as given by the Oxford Dictionary, defined it as 'the rudi-
mentary plant contained in the seed'.   However, the term 'embryo'
has been applied by most botanists not only to the seed plants but also
to the pteridophytes and bryophytes, which were even included with
the former in the Embryophyta on the basis of this feature (Campbell,
1940).   In the pteridophytes and bryophytes there is no seed, not even
a detectable resting state during development.   Furthermore, in these
two groups the development of the zygote was accepted as 'embryology'
because the zygote is an 'enclosed' structure, retained within the
archegonium, while the development of the spore into the gametangial
phase was excluded because it was 'free'.   The processes of development
of zygote and spore are not so different, and the enclosed or unenclosed
state of the body in question is not a sufficient criterion for discrimination.   There is thus a good case for extending the original concept and
applying the term to the early developmental stages of *all* plants,
including the algae (Wardlaw, 1955).

The stages of spore germination and subsequent development have
been described for a large number of red algae and it is obvious that
these do not conform to a single embryological pattern.   There is,
however, no agreement as to the number of different types which should
be recognised, the proposals ranging from three (Oltmanns, 1904;
Kylin, 1917a) to a very large number (Inoh, 1947).   The five categories
recognised by Chemin (1937) would appear to represent an appropriate
number of categories; these are as follows:

Nemalion type (Fig. 21, A-F).   The spore after attachment puts out
a prolongation into which all the contents pass and a transverse division
then occurs.   The cell formed from the prolongation functions as an
apical cell.   It divides to form a short filament of cells to which the
spore wall is attached at the lower end.   After some time, which varies
from species to species, a rhizoid is formed from the lowermost cell
but the number of cells in the germling may be considerable before this
occurs.   The *Nemalion* type of spore germination occurs in most genera
of the order Nemaliales.

Naccaria type (Fig. 21, M, N).   The spore after attachment usually
divides to form a number of cells of approximately the same size, although it may remain as a single cell.   The undivided spore or each of
the derivative cells then forms one or more protuberances which are then

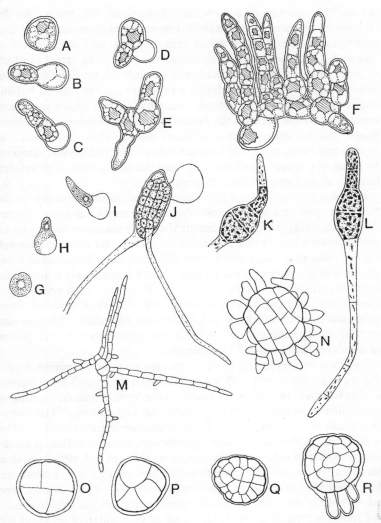

Fig. 21. *Embryology of Florideophyceae.* A, B, C, D, E, F, Nemalion multifidum (× 300); G, H, I. J, Gelidium pulchellum (G–I, × 200; J, × 325); K, L, Antithamnion plumula (× 350); M, Naccaria wiggii (× 200); N, Bonnemaisonia asparagoides (× 200); O, P, Q, R, Gastroclonium ovatum (× 150)

(A–F, *after Chester;* G–R, *after Chemin*)

cut off by transverse division walls. The cell or cells formed by the division of the spore or the undivided spore retain their contents. The protuberances function as apical cells dividing to form a series of filaments which soon give rise to lateral branches. The extent to which the spore does or does not divide and the subsequent development of filaments and branches is such that a prostrate mass of cells of variable outline and pattern is produced. Examples of the *Naccaria* type of spore germination occur in the Nemaliales, Cryptonemiales and Gigartinales and the unipolar germination of the so-called 'carpospores' of *Bangia* and *Porphyra* should be regarded as falling within this category. In these, the spore germinates and puts out a filamentous protrusion but the original spore retains its contents.

Gelidium type (Fig. 21, G-J). This is similar in some respects to the *Nemalion* type and is even included in the latter by some authors (Wardlaw, 1955). After attachment the spore contents emerge into a prolongation which is then separated by a transverse division although the subsequent development is very different from the *Nemalion* type. One end of the prolongation forms a rhizoid immediately whereas the opposite end segments to form a mass of cells within which one or more apical cells eventually differentiates. The *Gelidium* type of spore germination is best displayed in *Gelidium* and *Pterocladia* although patterns of spore germination of a similar type occur also in certain genera of the Cryptonemiales and Gigartinales.

Dumontia type (Fig. 21, O-R). This type of spore germination produces a relatively massive cell mass from which the erect frond is formed so that Chemin (1937) referred to this as the 'type moruléen' because of the similarity to the morula stage in animal embryology. The spores attach and then divide by a wall perpendicular to the substrate. The subsequent segmentation follows rapidly and, because of this, is somewhat difficult to determine but it appears that several further vertical divisions are followed by numerous divisions parallel to the substrate. In this way a hemispherical structure is produced. The *Dumontia* type occurs only in the Gigartinales, Cryptonemiales and Rhodymeniales.

Ceramium type (Fig. 21, K-L). This is very different from all the preceding types in that it represents bipolar germination. Following attachment, the spore develops two opposite primordia one of which forms a rhizoid while the other develops into an apical cell from which the erect frond is formed. The bipolar germination of the *Ceramium* type occurs exclusively in the members of the Ceramiales, of the Florideophyceae, although it occurs also in various genera of the Bangiophyceae. The 'monospores' of *Bangia* and *Porphyra* always germinate in this way while the 'carpospores' may germinate in a bipolar manner under certain environmental conditions.

The system put forward by Inoh (1947) uses further subdivisions based upon matters of greater detail. It would appear that the very large amount of material analysed so carefully and in such detail represents an overelaboration. It was pointed out by Chemin (1937) that the five types which he established showed well-marked differences but were in no way absolute in that various intermediate patterns of development were also known. Recent work indicates that the pattern of development of a spore is modified considerably by environmental conditions and the minor differences used by Inoh (1947) as the basis for his categories fall within the range of variability of a single entity. One most spectacular modification occurs in the so-called 'carpospores' of *Porphyra* and *Bangia* which may germinate in a unipolar manner under one photoregime but bipolar under another. Thus, the five category scheme proposed by Chemin should be regarded as a classification based on convenience, but nothing more than that.

Although there have been many descriptions of the morphological details of spore germination in red algae, there have been virtually no studies, either descriptive or experimental, of the many critical aspects which precede the actual segmentation of the spore. The factors governing the maturation and release of spores are a matter of conjecture. Tetrasporangia may be formed and tetraspores released within a matter of 12 to 14 days in some species of *Antithamnion* and *Callithamnion* whereas, in other red algae, the tetraspores may be formed within tetrasporangia but then not released for a considerable period under natural conditions. If released naturally, the tetraspores germinate although if released prematurely, by the application of mechanical pressure, adhesion and germination do not occur (Dixon, unpublished observations). Nothing is known of the maturation process involved. The adhesion of spores to a surface and the selective development of epiphytic species in relation to specific types of surface or particular host species is obviously of critical importance but, despite this, little information is available. In laboratory experiments (Linskens, 1966) it has been shown that spores of the Rhodophyta and of other algae fall into two categories, one of which develops preferentially on smooth glass surfaces while the other prefers a rough glass surface. This difference is independent of temperature over the range 13°C to 23°C, but it is worth noting that in green algae the gametes may belong to one category although the fused zygote belongs to the other.   Surface active agents applied to the glass surfaces eliminated the difference.   How these data can be applied in field studies is not clear.   The changes which occur in wall construction between the free spore and its adhesion are unknown although considerable progress has been made with similar studies of *Fucus* eggs and zygotes.   The construction of the spore wall is also of significance in respect of the amoeboid changes of shape and the movement of red algal spores over a surface, which have been well documented (Rosenvinge, 1927; Chemin, 1937).   Personal studies (Dixon, unpublished data) have substantiated these observations in a limited number of cases, although it is clear that neither changes of shape nor movement can be detected on most occasions.   Tetraspores of *Gelidium latifolium* and *G. crinale* were studied repeatedly over a period of years and amoeboid movements were detected on only three occasions but they were then extremely obvious and spectacular.   It is possible that these movements are associated with a maturation process in that the frequency of germination was much higher following the display of movement.   However, with other Rhodophyta, spores germinated on some occasions and not on others, but with no correlation with amoeboid movement.   Finally, as has been indicated elsewhere, little is known of the factors which control spore germination or the pattern of subsequent segmentation. *In situ* germination of spores has been reported on many occasions but the term has been applied both to the germination of a spore within its parent sporangium as well as to the germination of a spore which has been shed from its sporangium but then retained either within the thallus cavity, as in *Lemanea*, or in a relatively large flask-shaped pericarp, as in *Bonnemaisonia*.

Another phenomenon related to the control of germination and segmentation for which there is no explanation is the mutual interaction between segmenting spores which appears to occur frequently (Jones, 1956; Dixon, unpublished observations).   Thus, in culture, if spores are distributed sparsely and at a distance from one another for the purpose of facilitating observation, development may be slower or cease completely at an early stage.   This frequently involves the differentiation of the upright frond from the basal disc and it could well be that the failure to obtain complete development in culture is often related to this phenomenon.   Complete development does occur, on the other hand, if the spores are clustered and it frequently happens that under these conditions the basal discs formed from a number of spores coalesce.   When this occurs the coalesced discs function more as a single organism than as independent individuals and there is a considerable acceleration of growth rate and development. These effects are well shown by *Gracilaria verrucosa* (Jones, 1956) where the gametangial plants are dioecious. Two spores from each tetrasporangium produce spermatangial plants and two carpogonial so that coalescence probably involves spores of different genetical make-up. So far there have been no reports of coalescence between tetraspores and carpospores but such an occurrence would have several very significant biological implications and an experimental investigation would be well worthwhile.

There are many scattered reports of the development of spores *in situ*. This term has been applied in two very different ways. In one case, a spore is released from the sporangium in which it has been formed but fails to emerge from a surrounding protective structure or thallus cavity, while in the second case a spore begins to germinate while retained in the parent sporangium. The mode of development in either case may differ from that which occurs in a released spore. Nothing is known of the conditions which influence the occurrence of either type of *in situ* germination, although both types occur with a higher frequency in culture than under field conditions, while in the latter situation they are more likely to be found in drift specimens than in attached plants. Where tetraspores germinate *in situ*, they frequently merge with the tissues of the parent thallus. After some time, it is often difficult to distinguish, except by very careful examination, between the tissue of the diploid parent, producing tetrasporangia, and the haploid product formed by the germination of the tetraspore. This frequently gives rise to gametangia and it is possible that some records of the occurrence of various types of reproductive structure on the same plant are due to the occurrence of an *in situ* germling. In most cases where tetraspores germinate *in situ* from one to four of the spores begin to develop within the tetrasporangium and each gives rise to a distinct product. A more peculiar form of *in situ* germination of tetraspores has been reported in *Agardhiella, Champia, Cystoclonium, Gracilaria, Lomentaria* and *Pachymeniopsis* where the four spores in a tetrasporangium developed *in situ* to form a single compound individual (Fritsch, 1945; Schussnig, 1960).

Morphogenesis in red algal thalli

Morphological variation

The very considerable variation in thallus structure and form under different ecological, seasonal or geographical conditions is one of the major sources of difficulty in algal taxonomy and the repercussions of this problem extend to all aspects of phycology. Some of the early phycologists (e.g. W. H. Harvey) were exceptionally acute field observers and were well aware of the need for detailed knowledge of the variation in appearance of plants at different seasons or when growing under different conditions. Unfortunately much of the basic taxonomic work on the algae was undertaken by continental European botanists who were either unable or unwilling to collect personally and were, therefore, dependent on others for their material. Linnaeus is thought to have collected marine algae on only one occasion while Kützing and J. G. Agardh appear never to have worked in the field during the last 40 and 50 years of their lives, respectively, which in terms of publications were their most fruitful periods. As a result, the tendency to regard any morphological variant as a distinct taxon was subject to minimal challenge. Phycologists are at last beginning to take morphological variation into account but in many cases they refuse to consider anything other than the mature form of the organism. The comparison of organisms must be based not simply on the mature structure but rather on the pattern of development by which the mature form has come into being. For the algae where external form, despite the fact that it is subject to so much variation, is still the major criterion for taxonomic discrimination, a critical appreciation of morphogenesis and morphogenetic processes is indispensable. In the Rhodophyta very little information is available. In the Bangiophyceae, the thallus is of extreme simplicity and its morphology and form are of great flexibility. It can be shown that certain effects result from exposure to particular environmental or cultural conditions (Nichols and Lissant, 1967) but experimental investigation is difficult and, at the present time, the results achieved are not particularly relevant. Of the Florideophyceae, the encrusting thalli have also not proved tractable to investigation. Little is known as to how these encrusting thalli grow or of the ways by which the form of the thallus changes with age or is modified by environmental factors. Many crusts grow extremely slowly and it is very difficult to monitor growth in living material (Dixon, unpublished). On the other hand, the erect foliose thalli of the Florideophyceae have proved to be much more amenable to detailed field investigation and experimental study.

In such erect foliose thalli, it has been demonstrated (Dixon, 1963b, 1966a) that the external form of the thallus can be assessed in terms of: (*a*) the disposition of axes, (*b*) the shape of axes, and (*c*) the longevity of the erect thallus. These three variables are not, of course, entirely independent, but they provide a useful basis for the discussion of relevant data. The disposition of the axes of the thallus is determined by the position, number, and manner of formation of the filaments of unlimited growth. The formation of filaments of unlimited growth in many Florideophyceae is highly variable so that the thallus may be either simple or exhibit arrangements of branching of all degrees of complexity. In some genera, particularly in the Ceramiales, the formation of the filaments of unlimited growth is regular and rigidly controlled so that the disposition of the axes is characteristic of the taxon. In addition, the relative growth of these axes may be so co-ordinated that very elaborate thalli showing remarkable patterns of symmetry are produced. Despite the obvious importance of this feature, there is a surprisingly large number of genera for which information regarding the origin and disposition of the axes is fragmentary or completely lacking. The shape of individual axes is determined by the relative development of the filaments of unlimited growth and the surrounding filaments of limited growth. This may be considered most easily in those thalli where the latter filaments form a compact and uniform investment. The shape of such an axis is the result of the interaction between the growth in length, growth in breadth, and growth in depth. Growth of any filament is the result of increase in both cell size and number, so that the length of an axis is determined by the rate of division of the apical cell or cells of the filament or filaments of unlimited growth and the subsequent enlargement of the derivative segment cells, whilst the breadth and depth of an axis are related to the rates of division of the apical cells of the filaments of limited growth and the degree of enlargement of their segmentation products. In the Florideophyceae, the dimensions of the apical cells of the filaments of both limited and unlimited growth are relatively small and their segmentation products are of similar size when first formed. In view of the marked physiological differences which must exist between filaments of limited and unlimited growth, it is not surprising that the growth of the two types of filament should be differently affected by external conditions. Any factor which affects the growth of the filaments of limited growth, but not those of unlimited growth, or *vice versa*, or which has a differential effect on the growth of the filaments of limited growth in different directions, will have a profound effect on the shape of an axis. The subsequent enlargement of the derivative cells in both types of filament is usually very considerable; moreover, the degree of enlargement is usually extremely variable and is subject to considerable environmental modification. Although there are marked changes in the rates of division of the two types of apical cell, it is the variation in degree of enlargement of the segmentation products which has the greater effect on the variation in shape of the thallus. The effects of life span must also be considered because of changes in the appearance of plants between their first and subsequent years of growth. Studies of the perennation of thalli or parts of thalli have been somewhat neglected, despite the obvious importance of this feature. In many genera only the uppermost parts of a thallus disintegrate at the end of each growing season so that a considerable portion persists from one season to the next. The size of the surviving fragment, the number and disposition of new apices and the growth of the new axes all have a profound effect on the appearance of the thallus. In many cartilaginous species there is usually little degeneration of the apices, although these may be lost as a result of the formation of reproductive structures or by animal grazing. With these exceptions the original apices frequently function for a number of years. Increments are added to each axis during every growing season so that there are often marked differences in appearance between thalli of different age groups. In many genera the reproductive structures develop consistently in the tips of axes which tend to disintegrate following the maturation and release of the spores. Each year a new apical meristem may be regenerated from the stump of an old axis, as in *Furcellaria* (Austin, 1960a). Finally the breadth and depth of axes may also be changed as a result of perennation. Where the thallus survives for more than one growing season, the development of the filaments of limited growth in the second and succeeding years is usually negligible. There is thus very little change in breadth and depth of the axes except in certain special cases. In *Ahnfeltia*, for example, well-marked strata occur in the cortical tissues which have been interpreted as annual increments (Printz, 1926), although other investigators (Rosenvinge, 1931; Gregory, 1934) whilst accepting that the strata result from periodic bursts of growth deny that the latter occur annually. Similar strata occur in *Cryptonemia* (Feldmann, 1966).

These general principles of morphogenesis have been analysed in a number of examples (Dixon, 1960, 1963b, 1966a), such as *Grateloupia filicina*, *Pterocladia capillacea* and species of *Ceramium* and *Gelidium*, from various localities. In the case of *Pterocladia* it is obvious that there are differences in growth and behaviour in plants from areas outside the scope of the initial investigation. In New Zealand for instance the species appears to go through a similar cycle to that found in Britain, but with a phase difference of approximately six months, whereas in California, *P. capillacea* appears to behave very differently (Stewart, 1968).

It should be appreciated that such analyses are only the first step in an investigation of the way in which the thallus form comes into being. In order to understand fully the development of form it is necessary to have information on those factors in the environment which exert control over gross aspects of growth and development. Knaggs (1966a, b, 1967) has given analyses of the morphology of *Rhodochorton purpureum* collected from different ecological situations and attempted to relate this variation to specific factors in the environment. However, such ecological correlation is not proof that a particular response is produced by a specific factor in the environment although it does give an indication of possible lines for experimental investigation.

Recently, it has become increasingly clear that in the algae, as in other plants, a large number of processes, both structural and reproductive, are controlled outright by one or more factors, acting singly, together, or in sequence. For plants, the most obvious factor to be considered is light. It is now 50 years since it was shown that flowering in many Angiosperms was controlled by a sequence of light and dark periods.

The results obtained have been extremely confused and somewhat ambiguous in that the term 'photoperiodism' has been applied to a variety of situations in which at least three distinct phenomena can be delimited. The first of these is what might be termed a 'photosynthetic effect' in that different results are produced in relation to the amount of radiation received. A plant might respond to a long-day regime but not to a short-day regime because of the additional amount of energy obtained. The photosynthetic effect is sharply set off from the other two phenomena because of this energy relationship. The second process confused in the concept of photoperiodism involves a rhythm, or biological clock mechanism. In this, a cyclic process in the organism is kept in phase by the 24-hour cycle of light and dark. Cyclic phenomena related to circadian rhythms will continue in continuous light or continuous darkness although the cycle will drift in phase and the amplitude of the oscillation diminish with time under these conditions. True photoperiodism has been defined by Terborgh and Thimann (1964) as being characterised by: (*a*) induction, or continuation of the effect following transference to non-inducing conditions after a set time, and

(b) sensitivity to short breaks in either the light or dark periods. Most investigations of algae in which it is claimed that photoperiodism is involved have shown an effect under one set of conditions but not under another. In most cases the critical experimentation needed to distinguish between a photosynthetic effect, a circadian rhythm and true photoperiodism has not been undertaken.

The two principal topics of morphogenetic interest in the Rhodophyta, in terms of the morphological variation and the development of thallus form, concern (a) the conditions under which growth is initiated and ceases, and (b) the conditions responsible for seasonal changes in the growth pattern in terms of cell division and cell enlargement.

Considering first the problem of the conditions under which growth is initiated, there is at present almost no field data on which to base an experimental programme. The major difficulty with members of the Rhodophyta is that it is virtually impossible to say whether a meristem, if present, is active or inactive without destructive analysis. Such phenological studies as have been carried out have considered only reproductive structures rather than the initiation of growth and are based, largely, on sporadic observations without statistical analysis. *Pterocladia capillacea* is one species of the Florideophyceae in which it is possible to detect the earliest indications of the reactivation of the apical meristem, at least in British material. Each frond goes through an annual growth cycle in Britain and it is easy to detect the beginning of new growth even with the naked eye. The new growth is pale or even completely colourless in contrast to the purple-black coloration of the old axes of the preceding season on which it develops. In North Wales, observations of a single clonal population each year between 1950 and 1964 showed that the reactivation of the meristem could take place as early as the beginning of February in some years and as late as May or even early June in others. Such observations indicate clearly that a strict photoperiodic response is impossible and that a photosynthetic response is more likely. This is particularly interesting if one compares the British and the Californian material. In southern California, some 20° to the south of the previous observations, there is no indication of a growth cycle whatsoever and new growth appears to be initiated at all times of the year. In general, from various investigations in the laboratory, the initiation of development in a whole range of Florideophyceae (Dixon, unpublished) can be related to photosynthetic responses although the evidence available is far from complete. Such a conclusion is of obvious interest in relation to the ecological problems associated with perennation of basal fragments which are discussed elsewhere (see p. 211).

Conversely, Powell (1964) has observed that the initiation of the blade in *Constantinea subulifera* was induced by a critical daylength of 14 hours and that development was inhibited by longer daylengths or by a 15-minute lightbreak during shorter-day conditions. These observations are highly indicative of true photoperiodism as defined by the criteria of Terborgh and Thimann, the only instance so far reported of such an effect in relation to the initiation of growth in a member of the Rhodophyta.

The conditions affecting cell division and cell enlargement are now known for a number of Rhodophyta (see p. 100) but there have been few attempts so far to consider the differential effects on principal and lateral filaments such as would offer an explanation for the seasonal and environmental changes in form of many marine and freshwater red algae.   It was appreciated by many of the early phycologists (e.g. W. H. Harvey) that plants growing under conditions where light intensity was reduced, such as in turbid water, had very narrow axes.   The changes in shape of red algal thalli, particularly changes in breadth, which occur in relation to season, appear to show some relationship to the relative duration of light and dark periods but also to light intensity.   It would appear, therefore, that such changes are largely due to a photosynthetic effect and dependent on the total quantity of radiation received.

This survey of the gross changes in form of red algae and of the preliminary results which have been achieved with respect to the effects of the environment, in particular of alternating periods of light and dark, indicates the significance of such observations.   At the same time, the results which have been achieved offer little more than a tantalising glimpse of the potential behind this type of investigation.   The principal reason for the inadequate information obtained so far is that critical experimentation of a very high degree of sophistication is required. Maintaining an alga in a container in the laboratory is simply not sufficient for this purpose.   Only with careful control of *all* the conditions involved including light intensity, duration and quality, temperature and the various nutritional factors which bedevil all aspects of algal culture, will it be possible to state explicitly the conditions necessary for a particular growth response.   A further point, which applies to any study involving the use of algal material in culture, is that there must be constant reference, regarding conditions and the behaviour of material, between the laboratory and the field.   More precise definition of behaviour in the field will facilitate the design of critical experimentation for the investigation of such phenomena in the laboratory, while the results obtained in the laboratory must be shown to apply in the field.

Cell division and its control

Surprisingly little is known about the process of cell division in red algae, despite the considerable interest in cellular development of various structures such as the thallus or the carposporophyte. In general, it would appear that following nuclear division the cross wall develops centripetally by invagination. In most members of the Bangiophyceae this invagination closes completely although in genera of the Florideophyceae the invagination does not close completely, the cavity remaining as the site of formation of the pit connection (see p. 7).

In unicellular algae, the techniques of synchronous culture offer considerable possibilities for the investigation of biochemical and other factors controlling cell division (Hase, 1962). Such synchronously dividing cultures have been established for a number of unicellular algae, particularly members of the Chlorophyta, by subjecting homogeneous populations of young actively photosynthesising cells to light shock, alternating periods of light and dark or alternating temperature levels. For the red algae, such procedures do not appear to have the desired effect. In *Porphyridium purpureum* (as *P. cruentum*) alternating periods of light and dark increased the number of dividing cells although it did not result in complete synchronisation of cell division (Gantt and Conti, 1965). More recent studies (Gense *et al.*, 1969) have implied that complete synchrony had been obtained although not actually stating this, while Hoogenhout (1963) claimed to have obtained complete synchronisation in *P. aerugineum*, but this must be doubted in view of the yield variations reported.

Even if a system for synchronisation is perfected for the Rhodophyta it will be applicable only to a handful of genera and species. The major problem, analyses of cell division rates and the effects on these of internal and external factors in multicellular red algae, is still largely untouched. For larger, multicellular red algae, there have been various reports of mitotic periodicity (see p. 26) both in the field and under laboratory conditions although there appears to be an equal amount of evidence indicating no periodicity. There have been various studies involving the growth of thalli but almost without exception these have involved only the measurement of biomass or the examination of gross external features. The latter aspects are discussed in some detail on pp. 93-100. There have been surprisingly few measurements of the growth of red algae under any conditions (Dixon and Richardson, 1970) and in most of these no effort was made to discriminate between the effects of cell division or cell enlargement.

Members of the Florideophyceae are highly suitable for growth analyses in that, with certain special exceptions as in the Corallinaceae (see p. 57) and Delesseriaceae (p. 78), intercalary cell division does not occur. The growth of the filaments comprising the thallus results from the division of apical cells and the subsequent enlargement of the derivative products. The restriction of cell division in a filament to the apical cell means that the number of cells in a filament is a direct indication of the number of divisions of the apical cell. The rate of division of the apical cell can be measured simply by counting the number of cells. Studies of *Callithamnion roseum* (Konrad-Hawkins, 1964) showed that the apical cell of a principal axis divided 15 times in a five-day period but no attempt was made to influence this rate by altering external conditions. Studies of various other members of the Ceramiaceae indicate that the rates of cell division can be modified experimentally. Analyses of *Callithamnion tetricum* indicated that at a light intensity of 250 lux the number of cell divisions was related to daylength, following a lag period of two to three days at the beginning of the experiment (Dixon and Richardson, 1970). The investigation was far from complete but the evidence suggests a photosynthetic effect (see p. 97). A more sophisticated investigation of growth in *Pleonosporium dasyoides* (Murray, unpublished) shows that the rate of cell division is clearly under a photosynthetic control up to an intensity of 125 lux above which no increase in the rate of cell division occurred. In *Griffithsia pacifica* (Waaland and Cleland, 1972), under a photoregime of 16 hours illumination and 8 hours darkness per day, cell division shows a diurnal rhythm which persists for at least 7 cycles in continuous illumination and which can be 'reset'.

Cell enlargement

In genera of the Bangiophyceae it is obvious that some cell enlargement occurs following cell division but, because of the random occurrence of intercalary cell division in these thalli, it is not possible to offer any comments on the degree of cell enlargement or its control. Studies involving the statistical analysis of cell size (Dixon, unpublished) in species of *Porphyra* involved a considerable amount of effort with no

detectable results of any consequence.   The position in the Florideophyceae is very different and considerable progress has been made in
the study of cell enlargement in thalli of these organisms.   Excluding
those few genera for which intercalary cell division has been established,
the general picture of cell division and cell enlargement in the Florideophyceae may be summarised as a cyclic variation in size of apical cells
through enlargement and segmentation and the gradual enlargement
with age of the derivative products of apical cell division.

Apical cells.   There is enormous variation in shape, size, and mode of
division of apical cells throughout the Florideophyceae.   There may
even be considerable differences in these features between the apical
cells of the filaments of limited and unlimited growth, or between the
apical cells of the first, second, and subsequent orders of branches,
even in the same thallus.   The apical cells of all types of filament are
usually dome- or thimble-shaped, dividing by a series of divisions
parallel to the base to produce cylindrical segments.   The plane of
division may be inclined so that the segments formed are circular in
section but wedge-shaped in outline.   When inclined, the plane of
division may lie alternately to right and left, or follow a helical path
with three or more segments produced in the course of one rotation
through 360°.   In a few genera, such as *Gelidium*, the principal apical
cells of the filaments of unlimited growth are biconvex in shape,
dividing by a curved wall to form a concavo-convex segment cell and
retaining the biconvex shape of the apical cell.   In size, the smallest
apical cells are usually associated with filaments of limited growth and
measure as little as 5µm in length and 3-4 µm in diameter.   The largest
apical cells occur in species of such genera as *Griffithsia*, where they
may measure up to 800 µm in length with a diameter of 250 µm.   The
dimensions of all apical cells vary to a greater or lesser extent under
different environmental conditions.

As stated above, the apical cells undergo a cyclic variation in size
through enlargement and segmentation.   In general, the cyclic variation
applies only to length, the diameter remaining constant through each
successive division.   As indicated previously, cell division in the
Florideophyceae involves the formation of a cross wall by invagination.
In the apical cell this separates the portion which is to be cut off as
the derivative segment and the portion which is to remain as the apical
cell so that, in general, segmentation of the apical cell results in its
dimensions being minimal immediately after a division and maximal

immediately prior to it. The degree of variation in the length of the
apical cell varies from species to species depending on the relative size
of the portion cut off and the portion which remains.

Derivative cells. Except in certain special cases, as in the Corallinaceae
and Delesseriaceae, the cells cut off by an apical cell do not divide again
in the plane of division by which they originated. The derivative cells
can, however, divide in other planes to give rise to primordia which
either do not divide again, as in certain uncorticated species of *Polysiphonia* where they are represented by the pericentral cells, or function
as the apical cells of filaments of limited growth. Such lateral primordia
are not normally formed from segment cells other than immediately

Table 2. *Enlargement of axial cells in some Florideophyceae* (Dixon, 1970, 1971b)

	Size when formed, μm	Size when mature, μm	Increase in length	Increase in volume
Batrachospermum vagum	8×10×10	290× 50× 45	36×	815×
Ceramium rubrum	4×16×16	380×316×300	95×	35 000×
Ceramium echionotum	4×20×20	420×240×220	105×	14 000×
Antithamnion plumula	5× 7× 7	1500× 90× 90	300×	48 000×
Lemanea fluviatilis	8×12×12	8000× 80× 80	1000×	44 000×

behind the apical cell, although they may be formed through 'regeneration' in older thalli which have become damaged in some way. Thus,
in normal undamaged thalli, the derivative cells formed by the segmentation of the apical cell exhibit increase in length but no increase
in the number of cells between the base and apex. The two families,
Corallinaceae and Delesseriaceae, in which intercalary cell division has
been reported, exhibit a number of special features which have been
frequently misinterpreted. These intercalary cell divisions are not
random divisions occurring indiscriminately, but are the result of highly
organised processes akin to the division of normal apical cells. However
a cell is formed, by normal apical segmentation or through the intercalary cell division processes of the Corallinaceae or Delesseriaceae, the
processes of subsequent maturation involve enlargement.

The most significant feature of the Florideophyceae is that the
degree of enlargement can be enormous (Dixon, 1971b). Data on the
comparative sizes of axial cells are given in Table 2. Further detailed

analyses of cell enlargement in Florideophycean thalli have been made and several general features detected. The precise patterns of apical segmentation found in most Florideophycean thalli facilitate considerably such analyses in that the segments of any apical cell are formed in a regular sequence. Except in those members of the Delesseriaceae where intercalary cell divisions occur, the sequence of cells along a filament represents the age sequence, although the time scale involved may or may not be linear.

The first feature of common occurrence to be detected by such analyses concerns the asymmetric enlargement of individual axial cells (Dixon, 1971b). This can be demonstrated in any thallus in which axial

Table 3. *Enlargement of axial cells in some Florideophyceae* (Dixon, 1970, 1971b), *in terms of acropetal and basipetal components (all measurements in μm)*

Segment number	Batrachospermum vagum			Antithamnion plumula		
	Acropetal	Basipetal	Total length	Acropetal	Basipetal	Total length
5	3	5	8	2	3	5
20	4	15	19	7	21	28
40	7	116	123	15	87	102
60	8	164	172	18	186	204
100	10	230	240	40	940	980
150	10	280	290	50	1450	1500

cells give rise to pericentral cells. In such thalli, the pericentral cell or cells are formed when the axial cell is very young and the pit connection or connections are situated more or less at the median position along the length of the axial cell. The positions of these pit connections provide a useful 'marker' for monitoring subsequent changes in size with age. As the cell ages the length of the acropetal component increases only very slightly, the enlargement of the axial cell being almost entirely in the basipetal portion. The size ratio basipetal/acropetal is often of the order of 30 or more in mature thalli (Table 3). This asymmetric development is apparent in any uniaxial thallus where the occurrence of pericentral cells or other lateral structures is predominantly at the apical end of the axial cell in mature thalli (Figs. 14, B, 17). This predominance of basipetal enlargement appears to be an almost universal feature of the Florideophyceae although there are certain genera or species for which the concept requires some modification. One example is *Naccaria wiggii*, in which each axial cell gives rise to two filaments of limited growth arranged in an opposite manner. With age and enlargement, the two filaments remain on opposite sides of the axial cell although at maturity one is located at the acropetal and the other at the basipetal pole. Various species of *Polysiphonia* display an almost equal enlargement of the acropetal and basipetal poles of each axial cell so that the pit connections are approximately equidistant along the length (Rosenvinge, 1924). In addition, the pericentral cell associated with a tetrasporangium in this genus is usually attached to the lower half of the axial cell at maturity, although little is known of the steps leading to this arrangement. A similar disposition is found in some collections of *Griffithsia tenuis* with the lateral filaments, which are few in number but attached to the lower end of the parental cell. Unfortunately, this is another example for which the developmental stages are unknown. A further complication is that the basipetal occurrence of laterals in this species varies considerably in different geographical areas, being somewhat indistinct in Hawaiian collections (Abbott, 1946) but well developed in specimens from China and the West Indies.

The second feature of common occurrence is that there are patterns of cell enlargement within each Florideophycean thallus (Dixon, 1971b). If one considers the thallus in an uncorticated species of *Callithamnion* (Fig. 22, A), there is a single principal axial filament, each cell of which gives rise to a lateral filament. In the same way that axial cells, produced in a regular sequence, may be compared, so may the filaments arising from the axial cells, these also being formed in a highly regular manner so that the sequence of laterals also represents an age sequence. The apical cell of a lateral filament divides to produce segments which do not divide again in the plane of division by which they were formed. These segments enlarge and, at first, the enlargement is proportional to age. Before long the enlargement of the first-formed cells, that is, those nearest to the parent axis, diminishes, so that the largest cells in the mature lateral filament are not the oldest cells in proximity to the filament on which the lateral is borne but cells at some distance from the latter filament; thus, the cells at the base of a lateral are smaller than others which are younger (see Fig. 22, E). This reduction in cell size in a lateral at the point of attachment to the parent filament which diminishes with distance from the latter suggests that the phenomenon may result from the parent filament having some effect on cell enlargement within lateral filaments. A further deduction from this is that

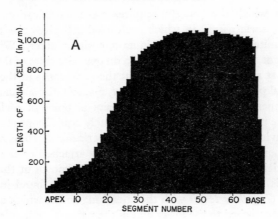

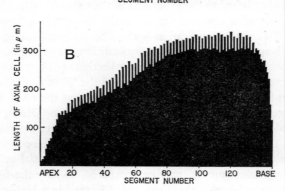

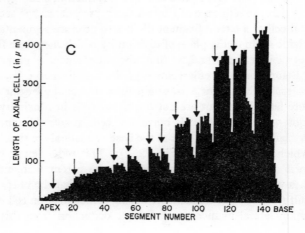

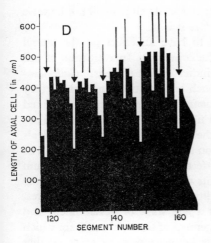

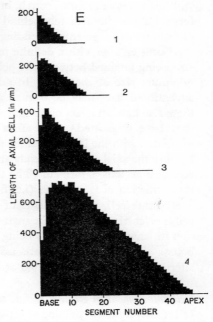

Fig. 22. *Cell enlargement in members of the Florideophyceae.* A, *histogram showing the length of every cell along a principal axis of* Callithamnion corymbosum; B, *histogram showing the length of every cell along a principal axis of* Ptilota plumosa; C, *histogram showing the length of every cell along a principal axis of* Ceramium rubrum (*the arrows indicate positions of 'dichotomies'*); D, *histogram showing the length of every cell along a portion of an axis of* Ceramium rubrum *on which both 'dichotomies' and adventitious lateral axes were present* (*the arrows indicate the positions of 'dichotomies' while the positions of adventitious axes are indicated by bars*); E, *histogram showing the lengths of every cell along a series of primary lateral axes of* Callithamnion corymbosum, *originating from the 10th (1), 20th (2), 30th (3) and 50th (4) cell of the principal axis*
(A–E, *after Dixon*)

the basal cell of a lateral filament or pericentral cell is small not because of *what* it is but rather because of *where* it is.

If the principal filament appears to have an effect on the lateral filament, is there any evidence for the converse? Investigation of cell enlargement patterns in many thalli indicates that there is. *Callithamnion* is not a suitable organism for such an investigation in that every cell bears a lateral filament. In *Ceramium* every axial cell is associated with filaments of limited growth, but filaments of unlimited growth are formed almost immediately behind the principal apical cell at intervals of from 5 to 14 segments depending on the species (Fig. 22, c, d). The axial cells enlarge with a definite pattern along an axis, the maximum size being attained between the points of origin of lateral axes and the minimum size at the axes. In addition to these initial lateral axes, adventitious axes may be formed at a late stage of development and these also have an effect on the enlargement of the axial cell from which they have originated. The effect is by no means as great as that associated with the laterals formed at the earliest stages of development because the axial cells are already almost fully developed by the time an adventitious axis is formed. The thallus of *Ptilota* is characterised by the formation of lateral axes (i.e. filaments of unlimited growth) from every other cell, with filaments of limited growth arising from every cell (i.e. 'cortication'). In the mature thallus there is a constant oscillation in length, alternate cells being longer or shorter than their neighbours. The shorter cells are associated with lateral axes and the alternate longer cells are devoid of the latter (Fig. 22, b).

Cell enlargement in relation to fine structure. The expansion of a plant cell results from the yielding of the cell wall to internal pressures (Heyn, 1931). The essential difficulty is that the evidence presented previously (p. 105) is indicative of differential cell expansion in different parts of the same thallus. As far as the Rhodophyta is concerned, little is known of osmotic phenomena (Guillard, 1962) and such evidence as there is gives no indication of differential pressures in parts of the same thallus. The information on wall structure in red algae presented previously (p. 3) may be summarised as the random occurrence of microfibrils of material akin to Cellulose II embedded in an amorphous matrix of mucilaginous material.

Although enlargement of a plant cell may be regarded as the gradual 'expansion' of the wall as it yields to internal pressure, the wall does not become progressively thinner as it increases in area. In many red algae, the wall thickness can increase as much as three times indicating that a considerable amount of new material is added to the wall during the process of enlargement. The way or ways by which this new material is added to the cell wall simultaneously with the increase in area has long been a problem. There are two ways by which it may occur. New material may be added to the wall on its inner surface or it may be intercalated throughout the thickness of the wall. The two methods are termed apposition and intussusception, respectively, and it should be appreciated that they are not mutually exclusive. The enlargement of both apical cells and of the derivative products of their division in terms of apposition and intussusception, have been investigated in only a very few examples, and neither in adequate detail nor in a sufficient number of organisms to answer many of the questions raised. With regard to apical cells, Strasburger (1882) demonstrated that the apical cell of *Bornetia secundiflora* had a wall which is markedly laminated and that as the cell increases in size new lamellae appeared to be laid down on the inner wall surface as the outermost lamellae ruptured; he obtained no evidence of intussusception. Kinzel (1956, 1960) examined both *B. secundiflora* and *Antithamnion cruciatum* and showed that there are several points of difference between the two members of the Ceramiaceae. The increase in the area of the wall takes place principally at the tip of the cell in *Bornetia*, whereas in *Antithamnion* the growth occurs predominantly in the basal part of the apical cell; this difference obviously needs further investigation, particularly in other Rhodophyta. More significantly, Kinzel disagreed with the earlier statement by Strasburger that there is no intussusception; according to him, the rupturing lamellae reported by Strasburger was merely a post-mortem lifting of the outer lamellae or cuticle and this created the impression of a conically tapering outer layer. Kinzel also claimed that both apposition and intussusception occur in the apical cell, with changes in the balance of the two processes, apposition dominating during division and intussusception during enlargement. Kinzel's statements have been confirmed by Priou (1962) in a reinvestigation of *Bornetia*. The elongation of cells in multicellular rhizoids has been ignored almost entirely, although in *Griffithsia pacifica* it has been shown that the apical cell of such a rhizoid elongates only at its tip (Waaland *et al.*, 1972).

With regard to the derivative cells formed by the division of apical cells, studies of these structures are few and none has been concerned with any aspect of the changes associated with the phenomenal enlargement which these derivative cells undergo during maturation. It has been claimed (Chadefaud, 1962; Priou, 1962), on the basis of histochemical tests and optical microscopy, that the wall structure in the filamentous Florideophyceae was of some complexity, being made up of several major components (see p. 3). Although the components appear to be well documented by Chadefaud and Priou, it is often difficult, if not impossible, to demonstrate the structures which they discuss. Ultra-thin sections examined under the electron microscope do not give any indication that the wall is built up in the manner described. In their discussion, neither Chadefaud nor Priou comment on the manner in which the structures which they describe allow the cell to enlarge or how such enlargement might take place.

Indications of the site or sites at which wall material is being added can be obtained by applying markers (such as charcoal grains) to the cell surfaces and then following their displacement by time lapse photography (Waaland et al., 1972). In this way the enlargement of derivative cells in Griffithsia pacifica has been shown to involve zones at both ends of the cell. The apically-directed growth zone is several times more active than that at the other end of the cell and the asymmetric enlargement mentioned previously results from this. The mode of enlargement found in G. pacifica has been termed 'bipolar band growth' and it differs from that found in cells of any other plant group.

Another problem related to the asymmetric enlargement of cells in different parts of the same thallus concerns the manner by which lateral primordia are formed by the further division of segments cut off by an apical cell. Considering the Florideophyceae as a whole, there appear to be two different ways by which lateral primordia may arise. These two ways may be exemplified by Batrachospermum and Lemanea. The segments formed by the transverse division of the apical cell in each of these genera give rise to a whorl of four primordia. The primordia in Batrachospermum are initiated by the development of very localised protuberances which are then cut off by invagination. The pattern of development in Lemanea is fundamentally the same although no protuberances are formed, the divisions being purely 'internal'. The crucial difference between these two genera may well be related to the ways in which new wall material is added: in Batrachospermum the addition of material in very localised areas results in protuberances, but apart from this the original segment does not increase in girth. The situation in Lemanea is very different in that the segment as a whole increases in girth before the 'internal' divisions take place and there are no protuberances. Thus it could be argued that new material is being added overall and not in localised patches as in *Batrachospermum*. This hypothesis regarding the localisation of protuberances is analogous to the suggestions put forward by Green (1960, 1962) to explain the location of lateral branch development in the green alga *Bryopsis*.

These differences in cell enlargement and segmentation exemplified by *Batrachospermum* and *Lemanea* occur in genera throughout the Florideophyceae. The type of 'internal' division met with in *Lemanea* has led Bugnon (1961) to postulate that this type of organisation is better interpreted as a parenchymatous structure rather than as pseudo-parenchyma, the view held by most phycologists. Bugnon's argument hinges on the extent to which the filamentous organisation is apparent in such thalli as *Batrachospermum* but obscured in *Lemanea*. It overlooks the exclusively apical cell division and polarity in both types. If the hypothesis presented previously relating to the localisation or general addition of material to dividing cells can be established in terms of fine structure, it will provide a real explanation for the apparently crucial difference between loose, obviously filamentous thalli and those thalli which are compact, albeit as filamentous in organisation as the less compact forms.

Polarity, apical dominance and regeneration

Two phenomena which, in higher plants, are both related to growth hormone behaviour are polarity and apical dominance, and it has been possible to demonstrate clearly the occurrence of these phenomena in red algae. So little is known of growth hormones in the Rhodophyta (p. 115) that the occurrence of polarity and dominance provides some of the strongest evidence for the existence of such hormones.

Polar phenomena in the Florideophyceae are widespread. The restriction of cell division to the apical cell in most entities within the group and the highly polarised pattern of cell division, even in those cases where cells other than the apical cell of a filament divide, provide some of the clearest indications of polarity. The pattern of cell elongation in a basipetal direction when there is no organised crystalline arrangement of the cellulose fibrils in the cell walls, such as determine the patterns of cell enlargement in green algae, is further evidence and it has been known for many years that cell maturation has a polar orientation (Child, 1916). One of the initial steps in the development of all groups of plants is the assumption of a spindle-shaped organisation, the so-called 'primitive spindle' (Bower, 1922), and this provides evidence for the establishment of polarity at a very early stage of growth. This primitive spindle may be related to the first division wall laid down in the germinating spore, as in members of the Ceramiales, although in those algae where a prostate crustose base develops first and the erect frond arises from this the indications of polarity may not be obvious at first sight. Every cell in a Florideophycean thallus is highly polarised and it seems difficult to conceive of a spore not being already polarised even in those cases where a basal crust is formed first. For carpospores in Florideophycean algae this is more understandable, because these are produced either by the apical cell or intercalary cells in an already polarised filament. It is more difficult to appreciate initial polarity in a developing tetraspore in that it has been shown (p. 125) that although the sporangial mother cell is an apical cell, a complete envelope develops within this prior to meiosis and cell division. Experimental investigation would be greatly facilitated if evidence were available on the ways by which polarity develops. Virtually nothing is known of the factors controlling the inception or the modification of polarity. It has been shown (Schechter, 1934, 1935) that the polarity of cells in *Griffithsia globulifera* (as *G. bornetiana*) may be reversed by moderate centrifugation and that, under the influence of an electrical current, rhizoids were formed towards the positive pole. It was claimed that chloroplasts also migrated towards the positive pole which seems somewhat curious in that, in the normal cell, the chloroplasts are more dense at the apical pole but somewhat sparse at the basal pole from which rhizoid development occurs. No explanation of any value has been proposed to explain these phenomena.

Apical dominance, in some form or another, can be demonstrated with ease in various red algae although explanation of such phenomena is still not possible. Apical dominance in the Florideophyceae is of several types. One of the simplest cases occurs in those thalli where lateral branches occur at every cell in an alternate or opposite manner. In the young undamaged state, the plant is built up on a distinct hierarchical system, with the growth of the primary axis greater than that of the secondary, growth of the secondary axis greater than that of the tertiary, and so on. In addition to differential growth between the orders of axes, there is also a dominating control so that if a major apex is lost or damaged, the apex of next lower order will take over its functions. The position in a member of the Dasyaceae must be much more complex because in thalli of such genera, dominance is accepted and then rejected after a set number of cell divisions to give a strictly sympodial organisation to the thallus.   In more complex thalli where morphological differentiation into filaments of limited and unlimited growth can be demonstrated, the process of limitation can be lifted if the apex of the filament of unlimited growth is damaged or removed. Such behaviour is approaching the phenomenon of regeneration, which will be treated next.

Regeneration phenomena in the Florideophyceae provide some of the clearest evidence for polarity (Fig. 23).   In some species, it is possible to isolate a single cell from a filament and this cell will produce a new thallus, with a new apical cell being formed from the apical pole and rhizoids forming from the basal pole.   This can be carried out experimentally with great ease in a species such as *Griffithsia corallinoides* where the cells may be separated one from another simply by putting a thallus into a bottle half-filled with seawater and shaking vigorously for a few minutes (Fig. 23, A).   Other filamentous red algae are much more coherent and cells can be isolated only by surgical treatment.

If a cell within a filament dies but without the thallus breaking, a situation often occurs which is the converse to that mentioned previously for *Griffithsia corallinoides*.   The cell on the apical side of the dead cell will produce a downgrowing rhizoid, while the cell on the basal side will produce a new apex.   If the cell size is small, then these regenerating apices and rhizoids will be external to the filament, whereas with larger cells, such as occur frequently in members of the Ceramiaceae, the regeneration can occur internally.   It would appear that the ingrowth occurs through the old pit connection although the evidence for this is not entirely explicit.   The ingrowing apices and rhizoids form a series of filaments which may fill completely the old cell lumen.   In some cases it would appear that this can occur even while the cytoplasmic contents of the receiving cell are still intact although no explanation for this latter phenomenon can be offered.   Stages in the formation of these 'internal cells' have been figured by many authors (Feldmann-Mazoyer, 1941; Drew, 1945; Funk, 1955) and described under a variety of terms, in most cases without any real understanding of the process which was occurring.   Sporangial regeneration (see pp. 126, 135) is analogous to this form of internal regeneration.   When the initial spore has been shed, a new apical cell is formed by what appears to be protrusion through the pit connection, the cutting off of this to form a new apical cell, and thence a new sporangium.

Regeneration in thalli of more elaborate morphology is a more complex procedure.   In young material of a species with well-differentiated

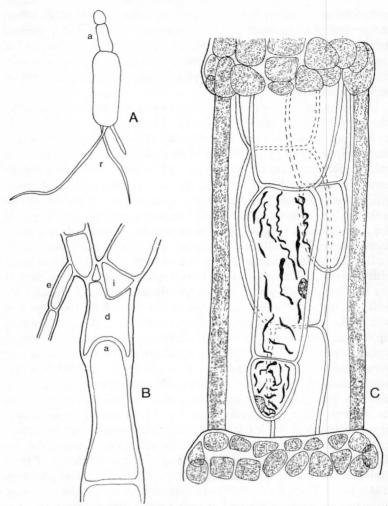

Fig. 23.   *Regeneration in Florideophyceae.* A, *isolated cell of* Griffi-thsia corallinoides *showing development of new apex* (a) *from apical pole and new rhizoids* (r) *from basal pole* (× 15); B, *effect of cell death* (d) *in* Pleonosporium borreri, *with development of internal* (i) *and external* (e) *downgrowing rhizoids and upward extension* (a) *of cell underneath* (× 90); C, *multiple development of internal rhizoids in an axial cell of* Ceramium tenuissimum (× 600)

(A, *after Chemin;* B, C, *after Feldmann-Mazoyer*)

axes, the general principles of hierarchical organisation will operate and a later, lateral axis take over the functions of the axis whose apex has been damaged or destroyed. In older material there may not be an axis remaining, or it may no longer be functional. New apices will regenerate usually through the conversion of the apical cell or cells of one or more filaments of limited growth into filaments of unlimited growth (Tobler, 1904, 1906). Where only filaments of limited growth are involved, these will regenerate rapidly through the re-induced divisions of adjacent apical cells. This can be seen most clearly in such flat membranaceous thalli as those of the Delesseriaceae. Perforations of the lamina will heal, through the reactivated division patterns of surrounding cells.

In members of the Bangiophyceae, the organisation of thalli is at a much lower level and there are by no means as many indications of polarity, either in the thalli themselves or in spores. In *Bangia* and *Porphyra*, for example, the polarity of the so-called 'carpospores' is influenced by the photoregime to which they are subjected, germination being bipolar under some conditions and unipolar under others. There are still, however, many instances of regeneration in members of this class and these could well be simply a reflection of their unspecialised condition.

The capacity to regenerate in all red algae is virtually unlimited and the ecological and biological implications of this are considerable. The ability of minute fragments which have overwintered, or survived through some other adverse period, often buried under a mass of sand or other debris, to produce new thalli as soon as conditions improve is one of the most significant reasons for the persistence of red algae, both in freshwater and the sea.

It will be obvious from this discussion that knowledge and understanding of the basic reasons for polarity, apical dominance and regeneration in red algae are very slight. In the most recent review of regeneration in lower plants (Müller-Stoll, 1965), the red algae are hardly mentioned. One can only take comfort from the fact that in higher plants, where these phenomena have been known for over two centuries, the level of understanding is but little higher.

Hormonal control mechanisms

The occurrence of patterns of cell division and enlargement in Florideophycean thalli implies the existence of some form of controlling mechanism as well as a transport system. As with other multicellular

E

algae, the occurrence of apical dominance and phototropic responses provide further circumstantial evidence for the occurrence in red algae of hormonal control of growth.   The gibberellins and auxins each play a major role in development in higher plants where they have been shown to influence the processes of cell division and cell enlargement. Their modes of action, even in higher plants, are still a matter of conjecture.

The lack of data on the occurrence of growth substances in red algae is shown by the fact that there is a single reference to the Rhodophyta in the most recent review of growth substances in the algae (Conrad and Saltman, 1962).   Since that time, there have been various reports of the occurrence and identification in red algae of gibberellins (Jennings and McComb, 1967) and auxins (Augier, 1965a, b; Schiewer and Libbert, 1965; Schiewer, 1967a, b; Schiewer et al., 1967).   The most recent study by Augier (1971) of Botryocladia botryoides suggested that there was evidence for five activators of growth and one inhibitor.   The application of gibberellins, auxin and kinetin to the Conchocelis phase of Porphyra (Kinoshita and Teramoto, 1958; Iwasaki, 1965) were said to have resulted in increased growth, with longer filaments.   Little else is known of the effects of applying growth substances to either axenic or non-axenic algal cultures.

It is obvious that a considerable amount of further work is required before the nature of the mechanisms controlling cell division and cell enlargement in Rhodophyta can be interpreted and, ultimately, controlled.   Even when these data are available, there remains the need to explain the organisation of the thallus in the Florideophyceae.   Differentiation between filaments in each axis and differentiation between axes may produce a highly regular, elaborately-branched thallus with a pronounced hierarchical organisation.   This hierarchical organisation is rigidly controlled in some Florideophyceae, such as the Ceramiaceae (l'Hardy-Halos, 1969b, 1971a, b), although subject to some environmental modification in other cases.   The method of control of this hierarchial organisation may be assumed to be of a hormonal nature although nothing is known of the mechanisms involved.   The hierarchical organisation in certain Ceramiaceae has been discussed in detail by l'Hardy-Halos (1969b, 1971a, b) and certain experiments related to growth, regeneration and differentiation performed.   On the basis of these, l'Hardy-Halos has postulated that the hierarchical organisation results from the interaction of two substances ('fluxes') moving towards the apex and towards the base of the thallus although admitting that, in addition, 'other factors of internal origin' must be involved also. Although incomplete, these investigations indicate the approach which will elucidate this problem, providing that analytical tools of greater sophistication can be developed.

Cell and tissue adhesions and fusions

Contacts between cells, filaments or axes resulting in adhesion or fusion occur much more frequently in red algae than one might appreciate from the literature. Examples occur both with respect to vegetative as well as to reproductive cells, although in most cases even the descriptive details are inadequate. There are virtually no physiological studies of these phenomena either in terms of the processes of adhesion or fusion, their consequences, or of the broader morphogenetical implications.

Considering first fusions which occur between vegetative cells in red algae, probably the best known is that involved in the formation in members of certain Florideophyceae of what are usually referred to as 'secondary pit connections'. At each division of an apical cell, the ingrowing transverse division wall fails to close completely and the aperture is closed eventually through the development of a plug (see p. 7). From the presence of such pit connections, which from what has been said occur only between cells of kindred origin, one can deduce the pattern of segmentation through which a morphological structure has developed. However, in addition to these primary pit connections, others may arise at a later date and connect cells which are not of kindred origin. These 'secondary pit connections' may develop between any two cells although they are most readily demonstrated in certain species of *Polysiphonia* and other rhodomelaceous genera where they occur between cells of adjacent rings of pericentral cells. In such cells, the nucleus divides and one derivative product passes into a protuberance which is then cut off from the mother cell. This protuberance is formed in the same manner as the apical cell of a lateral and a normal primary pit connection is laid down during its formation. There is then a series of co-ordinated movements, both of the cell which has given rise to the protruberance, and the cell to which the protuberance is going to fuse. The protuberance fuses with that cell and the primary pit connection laid down during the formation of the protuberance remains so that two cells which are not of kindred origin come to be associated by a pit connection. Connections of this sort are said (Fritsch, 1945) to be absent in members of the Ceramiaceae, although there is one subsequent report (Feldmann and Feldmann, 1948) of the occurrence of these structures in *Griffithsia*. Also, they have been said to be absent in members of the Nemaliales and there appear to be no subsequent reports of their detection in thalli of genera referred to that order, although they do occur in abundance in members of the Gelidiaceae which in the present text is not given independent ordinal status but referred to the Nemaliales. Secondary pit connections also appear to be absent in those members of the Cryptonemiales and Gigartinales where the thallus is of relatively simple organisation.

Secondary pit connections occur also in the Corallinaceae where they develop between cells of parallel filaments. Their manner of formation in this family is totally different from that discussed previously. Direct cellular communication develops between two cells through the dissolution of the walls at the point of contact. A plug-like pit connection, similar to that formed after cell division, is laid down in the channel of communication. One significant difference between the formation of secondary pit connections characteristic of the Corallinaceae and those found in the Rhodomelaceae is that nuclear transfer does not take place in the former.

Nothing is known of the fine structural or physiological aspects of those fusions which lead to the formation of secondary pit connections or of the physiological consequences of their formation. It has been argued (Fritsch, 1945) that secondary pit connections develop 'in response to special physiological requirements, especially longitudinal transmission of food material', but this is pure speculation with no supporting evidence.

Cell fusions which do not appear to be associated with the formation of secondary pit connections have been reported in some genera of both the Bangiophyceae and Florideophyceae. The filaments of the *Conchocelis* phases of *Bangia* and *Porphyra* are restricted to the interlamellar spaces when growing in shells. A dense mass of ramifying filaments is formed with abundant cell fusions, although no secondary pit connections are laid down. Apart from the chloroplasts, the mass of filaments has the appearance of a fungal growth and if the colour is lost or diminished it is very difficult to identify growths in a shell. It is highly probable that many of the reported shell-boring fossil 'fungi' (Pia, 1927) are *Conchocelis* or its equivalent. Cell fusions of this sort occur extensively in the Corallinaceae, both in the erect articulated thalli and the crusts, where the various minor variations in development have been discussed in detail by Cabioch (1972). Similar cell fusions have been reported in crustose thalli from families other than the Corallinaceae, such as *Cruoriopsis danica*, *C. gracilis*, *Rhodophysema elegans*, *R. georgii*, *R. minus*, and in a few erect thalli, such as *Ahnfeltia plicata* and *Rhodymenia palmata*. Cell fusions have also been reported in a few members of the Ceramiaceae, such as *Pleonosporium borreri*, *Bornetia secundiflora* and *Corynospora pedicellata* (l'Hardy-Halos, 1969a), which are of simple filamentous morphology. Nothing is known of any of the fine structural aspects of these cell fusions.

It should be realised that the organisation of many Florideophycean thalli depends upon the adhesion of the filaments of which they are composed. Under certain circumstances, this adhesion can break down. Nothing is known of the factors involved in either adhesion or break-down, although the latter occurs relatively frequently in cultures which are stale or ageing. When this happens the thallus is reduced to a mass of irregularly branched filaments which are very reminiscent of species of the *Acrochaetium/Rhodochorton* complex. Breakdown of the same sort occurs often in the green alga *Codium*, while in higher plant tissue cultures such breakdown is usually referred to as 'callus'.

In complex multicellular thalli, contact between multicellular axes may result in permanent adhesion. In general, it would appear that multicellular attachment structures are formed in response to a contact stimulus. Attachments may arise when the axes of two species come into contact, with rhizoids developing in some cases from both partici-pants, although it has been stated that such reciprocal rhizoid develop-ment only occurs when two axes of the same species are involved (Fritsch, 1945). More intimate contact can develop under certain circumstances and there is even some evidence to indicate that the formation of secondary pit connections can occur between axes of different species.

Various fusions are involved in the sexual reproductive process in red algae. The most obvious of these is in connection with gamete fusion. In Florideophyceae the fusion of gametes is well established (see p. 144) although nothing is known of the mechanism by which the male gamete is attracted to the carpogonium. That there is even such an attractive process can only be surmised from the fact that the process is far more efficient than it should be, on statistical grounds, bearing in mind that the female gamete is immobile and the male gamete non-motile. For the genera of the Bangiophyceae, the question of sex (see p. 174) is still too uncertain for further comment at this stage. A second, frequently more complicated, system of fusion occurs during the

development of the carposporophyte.   In a few cases the carposporophyte develops directly from the zygote (see p. 148) although in all
genera of the orders Cryptonemiales, Gigartinales, Rhodymeniales and
Ceramiales the zygote nucleus is transferred to another cell (see p. 147)
in a complicated series of developments.   The zygote nucleus or its
derivative is transferred by a process involving the fusion either of a
cell, or simply a protuberance, but the mechanism by which attraction
is effected is completely unknown.   The means of the attraction is likely
to be somewhat complex.   In genera of the Ceramiaceae, for example,
it is normal for a cell to be produced by the carpogonium after fertilisation and for this to be the means of transfer of the zygote nucleus or its
derivative product to the receptive auxiliary cell.   The auxiliary cell
is cut off from the support cell once fertilisation has taken place, and
on this basis it is relatively easy to develop hypotheses to account for
the attraction.   However, in certain cases (Dixon, 1964) the auxiliary
cell is *not* cut off from the support cell after fertilisation and the zygote
nucleus derivative is transferred directly to the undivided support cell.
In certain genera of the Cryptonemiales and Gigartinales, the primary
gonimoblast is extremely long and the receptive auxiliary cells are at a
considerable distance from the original carpogonium.   As with gamete
fusion, the odds against the auxiliary cells receiving the primary
gonimoblast on the basis of pure chance are very high and the frequency
with which this fusion is effected is far greater than one might expect
without some definite attractive mechanism.   It is probable that a genus
such as *Acrosymphyton* offers the best opportunity for detecting and
identifying this.   So far, all that can be said is that, on the basis of
histochemical tests, the primary gonimoblast and the auxiliary cells
contain concentrations of various enzymes far higher than those of
neighbouring vegetative cells.   In addition to the several methods of
fusion between the primary gonimoblast and the auxiliary cell, various
other types of fusion also occur involving cells adjacent to the developing carposporophyte.   In species of *Gelidium* for instance, the development of the carposporophyte may not involve any fusions whatsoever,
or the carpogonium may fuse with the hypogynous cell, or with neighbouring cells, or with both hypogynous cell and neighbouring cells.
The physiological explanations for the various developments are completely unknown.   Similar variations of fusion have been described for
a number of other members of the Florideophyceae and can involve
large numbers of cells, as in certain genera of the Corallinaceae.   For
the Bangiophyceae, there are no indications of any complex reproductive

developments as yet and no evidence of any fusions other than the most doubtful evidence for gamete fusion.

The final examples to be considered concern those adhesions which occur between spores or between the product of spore germination and the parental thallus on which it was formed. As has been discussed previously (p. 92), the spores of *Gracilaria* (Jones, 1956) develop better when clustered giving rise to basal discs which coalesce. Of the four spores from each tetrasporangium two give rise to spermatangial plants and two form carpogonial thalli so that the coalescence involves spores of differing genetical make-up. So far, there have been no reports of coalescence between haploid tetraspores and diploid carpospores. In most cases where tetraspores germinate *in situ* within the tetrasporangium, from one to four spores are involved and each gives rise to a new individual. In many cases, the germling thus formed falls away after a short time although if the tetrasporangium is embedded within the parent plant the germling may form a permanent attachment to the parent. When this happens, it is very difficult to distinguish between the tissue of the diploid parent and the haploid product of tetraspore germination. Gametangia are often formed on the latter and the germling resembles an adventitious lateral axis of the parent thallus. It is highly probable that some records of different types of reproductive structures occurring on the same thallus result from this phenomenon. A more peculiar situation occurs in those genera (see p. 93) where the four spores in a tetrasporangium germinate *in situ* to form a single compound individual. Just as the product of germination of a single tetraspore which has developed *in situ* can adhere firmly to the parent thallus a similar adhesion between parent thallus and the product of compound tetraspore germination *in situ* has been reported in *Pachymeniopsis* (Tokida and Yamamoto, 1965).

The descriptive details of fusions and adhesions such as have been outlined here are usually highly inadequate, physiological explanations for the phenomena are non-existent, and the morphogenetic implications are completely unknown.

Biological forms of Rhodophyta

It is now generally accepted that similar morphological patterns may occur in all algal divisions and that environmental modification of algal thalli is a widespread phenomenon. Is it possible, then, to use the form of the thallus as a means for the biological evaluation of the environment with the occurrence of different morphological types related to the particular habitat in which they occur? Such a concept had been developed for the Angiosperms by Raunkier (1934) and it has proved to be of considerable value in these respects for that group of organisms. All Angiosperms are of similar morphology in terms of the fundamental units of root, shoot and leaf so that the form of the whole is closely related to the methods of growth, longevity and the means of perennation. With algae, the early systems of classification of thalli (Oltmanns, 1905; Funk, 1927) were based entirely on static morphology and culminated in the scheme proposed by Gislen (1930). The Gislen system was applicable to all marine organisms, not merely algae, for which there were three categories, termed 'Crustida', 'Corallida' and 'Silvida', with several further subdivisions of each. Although widely used for descriptive purposes none of these systems was of much consequence as a means for the evaluation of the environment. A different approach, more analogous to the Raunkier scheme, was developed and elaborated by Feldmann (1937, 1951, 1966), based on the longevity of thalli, the means by which persistence was effected and the extent to which perennation occurred. The Feldmann system was obviously much more 'biological' than the preceding schemes and attempts have been made to relate the spectrum of thallus types with particular habitats (Ernst, 1958; Katada, 1963). Despite its usefulness, the Feldmann scheme suffers from the inescapable fact that there are in the algae no uniform morphological units such as characterise the Angiosperms. Because of the relatively enormous range of form and structure in the algae and the extent to which each part is capable of environmental modification, the Feldmann system is still far too imprecise to use for the purpose for which it was originally intended. The extent to which a system of this sort will ever be applicable depends largely upon the progress made with the morphogenetical understanding of thalli, not merely of the Rhodophyta, but of all divisions. Progress in this respect is still very slow and it would be unfortunate if the concept of 'biological' or 'life' forms in the algae should be abandoned without further consideration simply because of the general ignorance of the ways by which algal thalli grow and develop.

3

Reproduction

The most characteristic feature of reproduction in the Rhodophyta is that all reproductive bodies, whether spores or gametes, are devoid of flagella. The release of zoospores from fungal infections was probably the cause of various early reports of flagellated spores in red algae. It is very probable that the recent report of the detection of a flagellum in a cell of a spermatangial branch (Simon-Bichard-Breaud, 1971) of *Bonnemaisonia hamifera* is the consequence of a similar infection.

Discussions of reproduction in the Rhodophyta are often complicated by semantic problems, the terminology and nomenclature of the various types of spore which occur varying from author to author. This lack of clarity is largely a reflection of ignorance, particularly with the Bangiophyceae. Throughout the Rhodophyta, the nuclear phenomena are still uncertain in many sporangia whilst the products of germination of various spores are unknown. In addition, there are many incomplete observations which may or may not be of any significance. Because of such uncertainties, the role of a particular spore type may be a matter for conjecture, with very little real evidence on which to base an opinion. The present treatment will attempt to express the basic similarities and fundamental differences with the use of the least possible number of terms. One particular point which should be made clear concerns the use of the term 'sexual'. In many treatments of reproduction in the algae, the term is restricted to those structures associated with the process of syngamy. If one accepts the genetic definition of sex, meiosis is just as much a part of sex as syngamy, so that sporangia in which meiosis occurs should not be described as 'asexual' structures.

There are many significant differences in reproduction between the Bangiophyceae and Florideophyceae so that, as with vegetative structure, it is better to treat the two classes separately. At the present time, there are fewer uncertainties regarding reproductive processes in the

Florideophyceae than in the Bangiophyceae and it is better to treat the Florideophyceae before the latter group.

Reproduction in Florideophyceae

The Florideophyceae are characterised by a highly specialised type of oogamous reproduction in which the zygote is retained on the female gametophyte, giving rise to a complex post-fertilisation development known as the carposporophyte.   In addition, various types of sporangia occur in this class.   Their nomenclature is derived either from the number of spores produced or from the name of the genus for which a particular type is characteristic.   It is often inferred that the various categories are well defined and by comparison with the Bangiophyceae this is probably correct.   As will be shown intergrades occur between some of the accepted categories of sporangia so that these are not as well defined as has been assumed.

SPORANGIA AND SPORE PRODUCTION

Tetrasporangia

Tetrasporangia, each of which produces four non-motile spores, are the most widely distributed sporangia found in the Florideophyceae. Occasionally, tetrasporangia may develop in an intercalary position although, in most cases, they are formed terminally or laterally on fila-ments of limited growth.   The lateral tetrasporangium can, of course, be interpreted as the total conversion of a lateral single-celled filament of limited growth.   In simpler thalli, where the aggregation of filaments is minimal the tetrasporangia are free, whereas in more compact thalli the tetrasporangia are partially or completely embedded.   Tetraspo-rangia may be scattered irregularly over the surface of the thallus or restricted to particular areas, such as marginal proliferations (*Crypto-pleura*) or special axes (*Plocamium*, *Dasya*, etc.).   In certain genera of the Champiaceae the tetrasporangia are formed in sori, embedded in the superficial tissue, but as maturation and release occur the sori infold as sunken pits.   Tetrasporangia are restricted in many genera to nemathecia, which are superficial tissue masses formed as the result of resurgence of division in the apical cells of the filaments of limited growth in particular areas of the thallus.   As with other reproductive structures, the tetrasporangia in all Corallinaceae except *Sporolithon* are formed in conceptacles.   The tetrasporangia are formed on the floor of these and the tissue in which they are produced could be interpreted as a form of nemathecium which is subsequently enveloped by the growth of tissue on the flanks to form the flask-shaped conceptacle.

Tetrasporangia are of three types, cruciate, zonate, or tetrahedral, depending on the final arrangement of the four spores. The primordia are relatively small when first formed and usually uninucleate, even in thalli where the vegetative cells are multinucleate. Two reported exceptions to this generalisation are *Martensia* and *Nitophyllum*, where the primordia are initially multinucleate, all but one of the nuclei eventually degenerating. The occurrence of meiosis in tetrasporangia was first demonstrated by Yamanouchi (1906) in a species of *Polysiphonia* and there have been various subsequent investigations confirming that the nuclear division in these sporangia is meiotic. The possible absence of meiosis in some tetrasporangia is discussed later. The first meiotic division occurs when the primordium is still relatively small, but a rapid increase in size occurs whilst division is in progress. The relationship between nuclear division and cytoplasmic cleavage is a matter of some controversy. In tetrahedral tetrasporangia, the second nuclear division occurs prior to the correlated and simultaneous development of invagination furrows from the periphery inward. In most cruciate and some zonate tetrasporangia the first transverse cytoplasmic cleavage takes place before the second meiotic nuclear division, whereas in other zonate tetrasporangia the quadrinucleate condition is established before the first cytoplasmic cleavage. The first wall in cruciate tetrasporangia is transverse and this is followed by two more walls, one in each half, at right angles to the first and either at right angles to one another or in the same plane. Occasionally division of the two halves may take place by obliquely inclined walls to form a tetrasporangium with a highly irregular arrangement of spores.

Details of the fine structure of developing tetrasporangia are only just beginning to emerge (Chambers, 1966; Peyrière, 1969, 1970; Scott and Dixon, 1972a). One of the most interesting discoveries concerns the changes in wall structure during development and maturation. During early stages of tetrasporogenesis, a complete new wall is laid down within the existing wall of the mother cell, completely enclosing its contents. Each of the four spores, in turn, is enclosed by a wall which represents a third layer formed within the two layers then existing. The three layers are of somewhat different chemical composition in that the two outer react with both toluidine blue and PAS,* whereas

* Periodic acid Schiff reagent.

the walls of the spores do not. Another feature of interest is the presence of fibrous vesicles similar to those found in spermatangia although not as obvious or abundant as in the latter.

Surprisingly, little is known about the dehiscence of tetrasporangia. The spores may be released either by a regular splitting of the wall, or by a more irregular breakdown of the wall, or by the detachment of a lid from the apex of the sporangium. Following the release of spores from a tetrasporangium, the mother cell, lying beneath or to one side, may proliferate a new sporangium into the old empty wall. This process may occur more than once with the same sporangium.

Aberrant forms of tetrasporangia with more or fewer than the normal complement of spores have been reported in several genera. The occasional production of more than four spores in a tetrasporangium is rare and nothing is known of the cytology of these sporangia. The regular production of more than four spores in a sporangium is discussed in the following section. A reduction in the number of spores may also occur. In certain cases, the sporangia give rise to only one, two or three spores in an irregular manner and these are few in number compared with the normal tetrasporangia with which they are associated. In *Antithamnion elegans* (Feldmann-Mazoyer, 1941) and *Ptilotham-nionopsis lejolisea* (Dixon, 1971a) the single spore represents the release of the undivided cell contents but there is no evidence as to the viability of the body produced. The 'monospores' of *Nitophyllum*, reported by Svedelius (1914), are of a similar aspect to these although produced on the gametangial thallus. Where two spores are produced in a regular manner these are usually referred to as 'bisporangia', discussed in a later section, but these are often mixed with normal tetrasporangia and there is no clear-cut distinction between the categories. In *Litho-phyllum incrustans*, Bauch (1937) has reported that in some sporangia the products are not of equal size so that there is a single large spore associated with one or more minute bodies which probably represent degenerate spores. Conversely, in the Acrochaetiaceae it has been suggested that the tetrasporangia reported in members of that family represent 'divided monosporangia' (see p. 135).

There is considerable variation, from species to species, in the season at which tetrasporangia occur and even in the same species from different geographical areas. Phenological tables have been presented by various authors; the data presented (Table 4) are from Boney (1966).

Such tabulations are of considerable interest in relation to two separate problems although much of the data which have been obtained is unfortunately too imprecise. The two areas in which more exact information would prove extremely useful are, first, the biological/ ecological implications of the particular time at which a given spore type is produced in any species and, secondly, the morphogenetic problems associated with the various processes involved in the induction, maturation and release of spores and the effect on these of particular environmental conditions. In relation to the former, the time interval between initiation of the tetrasporangial primordium and the release of mature spores varies enormously from species to species. For some species, the process may be completed within 12-14 days for any

Table 4. *Times at which tetrasporangia have been reported from different localities*
(*after* Boney, 1966)

Species	Isle of Man	South Devon	Roscoff
Callithamnion tetragonum	A–W–Sp	Sp–Su	A
Ceramium flabelligerum	Sp	Sp–Su–A–W	Su
Hypoglossum woodwardii	A–W–Sp	Sp–Su	Sp–Su–A–W
Laurencia pinnatifida	W–Sp–Su	Sp–Su	W–Sp–Su
Lomentaria articulata	W–Sp	Sp–Su–A–W	Sp–Su–A–W
Plocamium cartilagineum	Sp–Su–A	A–W–Sp	Su–A–W

Sp, spring; Su, summer; A, autumn; W, winter

particular sporangium, with sporangia being produced continuously for a considerable period, whereas in *Furcellaria* on the North Wales coast (Austin, 1960b), the initials may be formed as early as April with meiosis occurring during a short period in early November and release of tetraspores taking place almost simultaneously throughout the plant in mid-December. Consequently, reports based simply on the occurrence of tetrasporangia upon a plant may bear little relationship to the time at which functional tetrasporangia are induced or released. The morphogenetic problems associated with spore production are of considerable significance and virtually nothing is known of the factors controlling the various steps in the process. Present information indicates that in any species, a particular spore type tends to occur approximately at the same time each year in any geographical area. Differences in time of development in any locality, from one year to the next, have been mentioned previously in connection with the initiation of growth and it is obvious from preliminary observations that analogous phenomena occur with tetrasporangia, so that the system is subject to some modification. Before comparisons between different geographic areas can become meaningful or preliminary suggestions made as to the nature of the control mechanism it is necessary to have much more precise information relating to the various steps between induction and the release of spores and the particular times at which these occur.

Quantitative estimates of the number of tetraspores released are extremely few. Boney (1960) has published estimates of spore production which indicate that there is considerable variation from species to species (Table 5). These data were obtained from collected material

Table 5. *Number of spores released per gram fresh weight per hour* (Boney, 1960)

Species	Number of spores
Ceramium ciliatum	438
Polysiphonia urceolata	354
Plumaria elegans	152
Polysiphonia lanosa	16
Polysiphonia nigrescens	5

maintained in dishes in the laboratory. There is no information available on behaviour in the field and it would be extremely advantageous to know whether there are any indications of daily, tidal or monthly fluctuations in spore release in view of the biological/ecological implications discussed previously. In many Florideophyceae, tetrasporangia are the only type of reproductive structure reported. As will be discussed later, this may indicate that there is an alternative morphological phase in the life history whose relationship to the tetrasporangial phase is still not appreciated. Another possibility is that the formation of tetraspores in a tetrasporangium does not necessarily imply the occurrence of meiosis. Information obtained by Suneson on *Lithophyllum corallinae* (1950) indicates that some sporangia may form four spores, but some only two, with meiosis occurring in both types whereas, in other sporangia, two spores are produced following mitotic nuclear division. This is the only instance of facultative apomeiosis in the Florideophyceae for which the supporting evidence is

reasonably sound.   The case of *Lomentaria orcadensis* is much quoted
but must be rejected.   In this species only tetrasporangiate plants are
known for western Europe.   The report of sexual plants (Lodge, 1948)
is based on a misidentification of the 'rough-water' plants of *L. clavellosa*,
whilst similar reports from Japan (Segawa, 1936) refer probably to
another species.   According to Svedelius (1937) the nuclear behaviour
in the tetrasporangial initial is of a special type, involving the migration
of the chromosomes into the nucleus, where mitotic division takes
place.   Reinvestigation by Magne (1964a) shows that meiosis is prefectly
normal and that the claims of Svedelius are based on misinterpretation
of granules within the nucleolus.   The situation in this species is still
not fully understood, in that Magne obtained chromosome counts
during diakinesis of $n = 10$ and $n = 20$ in material from Brest and the
Baie de Morlaix (Finistère, France) respectively, but, unfortunately, no
mitotic counts for comparison.   Magne (1959) himself, has also suggested that apomeiosis may occur in the tetrasporangia of *Rhodymenia
palmata*, for which these are the only reproductive structures reported.
This investigator reported counts of 14 chromosomes in all cells of
the thallus but observed 14 bivalents in the sporangial initial, the
nuclei of which appear to undergo normal meiosis.   In view of the
reports of counts of 21 chromosomes in the same species by another
investigator it may well be that there are complications in this species
resulting from polyploidy and a complete explanation of these data is
still not possible.   In conclusion it can be stated that although there
is some evidence suggestive of apomeiosis, no absolute proof of its
occurrence has been obtained other than as a facultative phenomenon
in *Lithophyllum*.   Furthermore, in view of the many instances where
what are now known to be different phases of the one life history were
once regarded as distinct species or even genera and placed in widely
divergent families or even orders (see p. 192), the argument that the
presence of tetrasporangia as the only *known* form of reproductive
structure *must* be proof of apomeiosis has to be rejected.

Polysporangia and parasporangia

Sporangia containing more than four spores have been reported in
various genera of the Florideophyceae.   The terms 'polysporangium'
and 'parasporangium' have been applied to these structures somewhat
indiscriminately and the situation is extremely confused.   Schiller
(1913) attempted to differentiate between the two types of sporangium
but his system proved to be inapplicable.   As a result of her cytological

studies, Drew (1937, 1939) has suggested that in the Florideophyceae the sporangia containing more than four spores are of two categories:

(a) Those sporangia which are homologous with tetrasporangia, in that meiosis occurs in the course of spore formation and the structures occur in the organism in place of tetrasporangia.

(b) Those which are not homologous with tetrasporangia, in that meiosis does not take place during spore formation and the structures do not occur in the organism in place of tetrasporangia.

She suggested that the term 'polysporangium' should be applied to the first type of sporangium and that 'parasporangium' should be retained for the second. It is unfortunate that there is still insufficient cytological evidence available to permit the complete application of this system and that interpretations are based usually on the secondary characteristic.

Sporangia which appear to fall within Drew's concept of polysporangia occur in the genera *Pleonosporium*, *Aristothamnion* and *Tiffaniella* (Fig. 24, c), as well as in many species of *Spermothamnion*. It is often stated that tetrasporangia are unknown in those genera and species. Funk (1927) once reported the occurrence of tetrasporangia in *Pleonosporium borreri* whilst Simons (1960) has described the occurrence of tetrasporangia in *Aristothamnion purpuriferum* on the same plants as carpogonia, with polysporangia on separate thalli. The most detailed cytological investigation of polysporangia is that by Drew (1937) who showed that in *Tiffaniella snyderae* (as *Spermothamnion snyderae*), the sporangial primordia are initially multinucleate, containing from two to nine nuclei. None disintegrates, but all undergo meiosis to produce from 8 to 32 daughter nuclei. These become arranged around the periphery of the sporangium which then divides by invagination furrows. The furrows are not arranged always in a regular pattern, but are sometimes so irregular that spores of different size may result. The spores are usually uninucleate, although binucleate spores are formed occasionally, whilst in one very irregular sporangium a spore with four nuclei was observed. These observations are very similar to those of Miranda (1932) on *Pleonosporium borreri*. *Coeloseira* and *Gonimophyllum skottsbergii* are other instances of Florideophyceae where sporangia which may be interpreted as polysporangia occur, with tetrasporangia absent. These polysporangia appear to differ in certain respects from those of *Spermothamnion* or *Pleonosporium* although the information available is by no means as complete. The development of the polysporangia in two species of *Coeloseira* (Hollenberg, 1940)

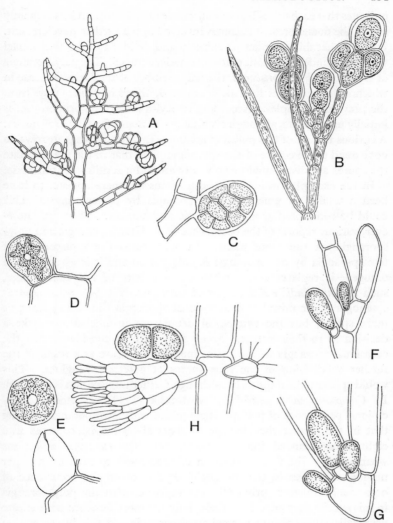

Fig. 24. *Sporangia in Florideophyceae.* A, *parasporangia,* Calli-thamnion hookeri (×70); B, *seirosporangia,* Seirospora seirosperma (×235); C, *polysporangia,* Pleonosporium vancouverianum (×250); D, E, *monosporangia,* Acrochaetium virgatulum (×1000), E, *showing mode of spore release;* F, G, *monosporangia,* Corynospora pedicellata (×40); H, *bisporangia,* Crouania attenuata (×250)

(A, B, *after Rosenvinge;* C, *after Kylin;* D, E, *after Boney;* F, G, H, *after Feldmann-Mazoyer*) indicates that each arises from a multinucleate primordium in which only one of the nuclei present becomes involved in the further development. It is said that this one nucleus enlarges and divides to form four nuclei which *then* undergo meiosis to form 16 derived nuclei. The polysporangia of *Gonimophyllum skottsbergii,* the only member of the Delesseriaceae in which such structures have been reported, differs in various ways from the previous examples. According to Kylin (1924) the primordium is initially uninucleate, although eventually it contains from 30 to 50 nuclei. A curious feature of this polysporangium is that the cleavage furrows do not extend to the centre of the sporangium, so that in the mature state the spores are distributed peripherally around a residual central mass.

In the examples which have been discussed, there appears to have been a total replacement of tetrasporangia by polysporangia. This could be considered as a phylogenetic replacement. There are, however, various reports of the occurrence of both tetrasporangia and polysporangia on the same plant. In such cases the replacement of tetrasporangia by polysporangia is only partial and it is essentially an ontogenetic replacement. In certain members of the Champiaceae, such as *Chylocladia reflexa, Gastroclonium coulteri and G. ovatum,* tetrasporangia occur mixed with occasional sporangia containing eight or more spores, but the cytological details are insufficient to make a decision as to their nature. Some early authors considered that the occasional sporangia with eight or more spores arose as a result of the further subdivision of the four spores of a tetrasporangium. This would appear to be improbable although further information is required. In *Compsothamnion thuyoides,* Westbrook (1930) reported the occasional occurrence of polysporangia mixed with tetrasporangia, stating that in the former the cleavage of the cell sporangial contents into eight spores occurred simultaneously and that there were no indications whatsoever of the further division of tetraspores to produce a larger number of spores in the sporangium. She noted the occurrence of occasional binucleate primordia and suggested that the polysporangia in this species may arise from these, with the more frequent uninucleate primordia giving rise to normal tetrasporangia. If this interpretation is correct, the occasional polysporangia mixed with tetrasporangia must be considered as multiple tetrasporangia.

It has been argued (Church, 1919) that polysporangia represent a relic of the ancestral type of sporangium of the Florideophyceae, but this is unlikely as polysporangia occur predominantly in the Ceramiales, generally acknowledged as the most advanced group.

In addition to these relatively clear-cut instances where a definite sporangium homologous with a tetrasporangium gives rise to more than four spores there are numerous other many-spored structures which agree with Drew's concept of the parasporangium. As far as their position is concerned, these parasporangia may or may not arise in a position comparable to that of tetrasporangia. Their origin is essentially the result of two processes, cellular proliferation and separation, which may be initiated in a single terminal cell or in a group of cells. In *Plumaria* and *Callithamnion hookeri* (Fig. 24, A) the occurrence of cells intermediate in appearance between vegetative cells and paraspores (Rosenvinge, 1924) supports this opinion. The idea that in these two taxa the parasporangium may be interpreted as a modified tetrasporangium is simply a result of the latter arising in the same position. In *Ceramium* and *Antithamnion plumula* (Schiller, 1913) the parasporangia are never found in the same position as tetrasporangia so that in these examples such an interpretation is impossible. Cytological details are known only for the parasporangia of *Plumaria elegans* (Drew, 1939), where the nuclear divisions are entirely mitotic. The number of spores produced by each parasporangium varies considerably, from six or seven in the smallest parasporangia of *Plumaria elegans* to several hundred in certain species of *Ceramium*. The spores appear to be exclusively uninucleate. In the latter genus, there has been confusion between parasporangia and the gall-like proliferations which occur so frequently in this genus. It is not entirely clear whether the two phenomena are related; both involve cellular proliferation, but there is no subsequent separation of cells in the galls, such as occurs in the parasporangia. In general, the cells which have separated are shed from the retaining membrane at release and germinate in the normal manner for spores in each genus (Schiller, 1913).

The 'polysporangia' reported by Howe (1914) in *Acrochaetium polysporum* are of very peculiar appearance bearing little resemblance to any other sporangia, being produced in catenate chains both in the prostrate and erect systems whilst the polyspores were discharged in a mass. It would appear most likely that these structures were cells infected with some fungal parasite.

Bisporangia

The occurrence of sporangia containing only two spores has been reported in many Florideophyceae, particularly in members of the Corallinaceae and Ceramiaceae (Fig. 24, H). Bisporangia occur either as the only form of reproduction or mixed with tetrasporangia. There are no reports of their occurrence on either carpogonial or spermatangial plants. The general impression (Fritsch, 1945) is that bisporangia are homologous with tetrasporangia. This impression may have been strengthened by the delay which can occur between the first transverse cleavage and the subsequent divisions in many cruciate (and some zonate) tetrasporangia so that two-celled structures often appear in quantity on what are actually tetrasporangial thalli. Whether viable spores can be released from these incompletely developed tetrasporangia is the crucial question, for which there is no answer at the present time.

It has been shown (Bauch, 1937) that bisporangia are of two types, the mature spores being either uninucleate or binucleate at the time of release. Subsequently Suneson (1950), in a cytological investigation of *Lithophyllum litorale* and *L. corallinae*, showed that in the former, for which plants bearing only bisporangia are known, meiosis did not occur in these reproductive structures. Two nuclei are formed by mitosis and these pass to the two spores, each of which is uninucleate at maturity. In *L. corallinae*, on the other hand, gametangial and bisporangial plants are known from Swedish waters, although tetrasporangial plants have been reported in the Mediterranean. Suneson showed that in Swedith material three types of sporangia were produced—quadrinucleate tetrasporangia, quadrinucleate bisporangia, and binucleate bisporangia, with meiosis occurring in the first two types, but not the last. On the basis of these observations the bisporangia in this species are clearly homologous with tetrasporangia. However, the contradictory reports of the occurrence of bisporangia in various ways, in different species, and in different geographical areas indicate that caution is needed with regard to their interpretation. In *Crouania attenuata*, both in the North Pacific and North Atlantic oceans, there is good evidence to suggest that bisporangia are formed during the winter months with tetrasporangia produced on the same plants during the summer. Nothing is known of the cytology of the two types of sporangium in this species or of the possible factor or factors in the environment controlling the two patterns of development.

Monosporangia

Various types of sporangium may be grouped together under this designation, the one common feature being that a single spore is released from each (Fig. 24, F, G).

Monosporangia occur in various genera of the Nemaliales as spherical or oblong bodies, usually distinguished from spermatangia by their pigmented contents and larger size. Distinction between monosporangia and spermatangia in these Nemaliales can pose problems, in that the two types of structure may occur together whilst alleged transitions between them have been reported in both *Batrachospermum* and *Scinaia*. In *Helminthocladia*, certain structures have been identified as monosporangia by some workers or as arrested spermatangial branches by others. Monosporangia have been reported in many species of *Acrochaetium*, whereas in the closely related *Rhodochorton* there appears to be only a single record of their occurrence. In *Acrochaetium*, the monosporangia are sessile, occurring either singly or in clusters at the ends of filaments, as in *Helminthora* or *Batrachospermum*. The monosporangia of the latter genus are extremely common. They may occur on both the upright gametangial frond and the prostrate *Chantransia*-stage, although in many species they are restricted to the latter. By comparison, in the closely related genus *Lemanea*, there appears to be only a single record of their occurrence, on the *Chantransia*-stage. The monosporangia in *Scinaia* are formed in young parts of the thallus from small pigmented cells which are intermixed with the larger vacuolate superficial cells. The monospores of the Nemaliales are all uninucleate and are liberated by rupture of the sporangial wall. Regeneration of the sporangium occurs by proliferation from the cell below and this can occur many times in one sporangium. Each time, the empty sporangial wall remains, with a varying degree of distortion, and it has been indicated that this can happen as many as 11 to 14 times in *Acrochaetium virgatulum*. It has been suggested (Kniep, 1928; Fritsch, 1942) that in species of *Acrochaetium* the tetrasporangia which may occur are 'divided monosporangia'. This suggestion stems from reports of the occurrence of small numbers of tetrasporangia mixed with monosporangia. A more likely explanation (Boney, 1967) is that when a monospore of *A. virgatulum* germinates *in situ* (see p. 126) the early stages of development may show considerable superficial similarities to a tetrasporangium.

The monosporangia of *Corynospora* (= *Monospora*) are very different from those found in the Nemaliales. In this genus the monosporangia are multinucleate structures formed by the direct modification of the apical cell of a two-celled branchlet. These two-celled branchlets are adventitious structures formed in the 'axil' between main and lateral axes. It has been argued (Schiller, 1913), that the two-celled arrangement of monosporangium and stalk cell represents a modified bisporangium. This is improbable in that the sporangium does not shed a spore as would be expected if this were a modified bisporangium, but is shed as a whole. Also, under estuarine conditions, the two-celled branchlet is frequently reduced to a single cell. Cytological examination has given some suggestions that the nuclei within the sporangium undergo meiosis; this would seem highly improbable judging from the spore behaviour but the indications of diakinesis are as clear as in any red algal meiosis. The 'monosporangia' of *Nitophyllum* are also multinucleate (see p. 125), but little is known of these structures.

A further type of monosporangium occurs in *Ahnfeltia*. Here, the monosporangia are formed in superficial colourless nemathecia. These nemathecia are so distinct in *A. plicata* that they were once considered as an independent 'parasitic' Florideophycean alga, *Sterrocolax decipiens*. These structures are discussed later in relation to carposporophyte development and life histories in the Phyllophoraceae (p. 199).

The production of monosporangia in terminal chains characterises certain species of the genus *Seirospora*, *Dohrniella neapolitana* and *Gibsmithia hawaiiensis*. These structures are frequently referred to as 'seirosporangia', but are better treated simply as a form of mono-sporangium (Fig. 24, B). They are formed in branched or unbranched chains at the apical extremities of filaments with progressive transformation of vegetative cells. The latter are uninucleate as are also the derivative sporangia. The degree of packing between the sporangia varies considerably. In some species of *Seiropora* the sporangia are ovoid, in loose chains, whereas in *Dohrniella* they are more tightly aggregated and cylindrical in shape. The walls of these sporangia are thicker than those of the vegetative cells from which they have arisen. Germination of the monospores of *Seiropora* may take place in either of two ways. The spore may be shed from the thickened sporangial wall, or the entire sporangium may be shed, as a whole, but with the contents emerging at germination. It is possible that the latter process is a consequence of experimental handling in that it does not appear ever to have been reported in 'wild' material. The formation of mono-sporangia in chains by the transformation of vegetative cells is, in fact, little different from the process of formation of parasporangia in *Plumaria elegans* (see p. 133).

The occurrence of monosporangia has been detected in cultural studies (see p. 203) of a wide range of organisms where they have not been reported in the field, including *Pseudogloiophloea confusa* (Nemaliales) and *Halymenia floresia*, *Thuretellopsis peggiana* and *Pikea californica*

(*Cryptonemiales*). These monosporangia are little-differentiated and the process appears to consist of little more than the release from a cell of its contents. The cells from which the contents emerge frequently develop in pairs, on either side of elongate filaments, and there are some cases where the structures are formed but with no evidence of any release. This process is relatively common in members of the Bangiophyceae and it is likely to be detected more frequently in members of the Florideophyceae in the future as a consequence of more sophisticated techniques of culture and more detailed, careful observations. The most critical feature of these monosporangia is that they have been detected only in laboratory culture and there is still no evidence for their occurrence in the field.

GAMETANGIA AND GAMETE FORMATION

The significance of sexuality in the Florideophyceae and the nature of the gamete-producing reproductive organs was established by Bornet and Thuret in the latter half of the nineteenth century. In a series of superb studies of a range of genera collected on the Channel and Atlantic coasts of France these investigators first demonstrated the structure and disposition of male and female gametangia and the process of fertilisation (Bornet and Thuret, 1867, 1876, 1880; Thuret and Bornet, 1878).

The carpogonium

The carpogonium is the female organ. It consists of an inflated base which contains the reproductive nucleus and an elongate process, the trichogyne, which functions as the receptive area for male gametes.

The carpogonium is formed by the transformation of the apical cell of a filament which may contain a specified number of cells, often three or four, and arise terminally or laterally as an adventitious and specialised structure. In such cases, the cells are often very different in size and shape from neighbouring vegetative cells and they are usually devoid of chloroplasts. The filament associated with the carpogonium is usually referred to as the carpogonial branch. In many of the lower Florideophyceae this is by no means as distinct as it is, for instance, in the Ceramiales, so that there are many problems with interpretation. The degree of specialisation of the cells associated with the carpogonium is variable in *Batrachospermum* and many other members of the

Nemaliales and the cells are not sharply distinguished from vegetative cells. Chloroplasts diminish in number from the normal vegetative cells to the carpogonium although there are usually some present even in the latter cell in such genera. The absence of a sudden and dramatic change in appearance between the vegetative cells and the cells associated with the carpogonium, such as occurs in the Ceramiales, means that in many of the Nemaliales and lower Florideophyceae it is impossible to define exactly what is meant by the carpogonial branch. The carpogonium in certain genera is a sessile structure, without any associated structure which can be defined as a carpogonial branch. Such sessile carpogonia are formed by the transformation of a lateral apical cell, that is, the apical cell of a single-celled lateral branch. The transformation may be effected in certain cases before the protuberance which gives rise to such a lateral apical cell has separated completely from the mother cell so that the resulting carpogonium is then an intercalary structure.

The division of the apical cell ceases and it remains for some time in this condition. The tip of the apical cell then develops a protuberance which inflates rapidly to form the trichogyne. There are very few observations on this process of transformation even with optical microscopy and it is clearly a situation where details of fine structure would be of considerable interest. Technical difficulty is the principal reason for lack of information, although in some of the more diffuse thalli of the Ceramiales the carpogonial primordium may be detected and its development followed with relative ease compared with those genera of the Cryptonemiales and Gigartinales where the thallus is compact. Observations on species of *Antithamnion* and *Pleonosporium* indicate that the apical cell remains for about six hours after the last division before the development of the trichogyne commences and that this is completed in about eight to twelve hours. Whether this time sequence applies in other genera is unknown but the infrequency with which developmental stages of the trichogyne are found suggests that it must be a very rapid process.

The size, shape and structure of the trichogyne and its persistence vary enormously. In many Gigartinales the trichogyne is a short cylinder about 5 μm in length whereas in many Ceramiales it is relatively long, achieving a length in excess of 100 μm. The trichogyne is usually cylindrical in shape although it may be coiled or twisted, particularly in compact thalli, so that the convolutions may simply be a consequence of the difficulties of emergence. The relative volumes of the basal portion and the trichogyne are about equal in most Florideophyceae. There is one notable exception to this, in *Batrachospermum*, where there are several, very characteristic types of trichogyne, in all of which the volume of the trichogyne is many times that of the basal portion. Despite their immense size, these trichogynes are remarkably persistent, often surviving until post-fertilisation development is complete although there may be some distortion during this time. In most Florideophyceae the trichogyne is a relatively ephemeral structure, whose survival must be measured in hours. The wall of the trichogyne is often quite thick and the frequency with which detritus particles are found adhering to it indicates that it is somewhat mucilaginous. The reproductive nucleus is always located in the basal portion but there is some disagreement as to the occurrence of a second nucleus in the trichogyne. Even where this has been reported, it is said to degenerate rapidly. Svedelius (1908) reported the occurrence of several nuclei in the carpogonium of *Martensia*, but the vegetative cells are multinucleate and the nuclei reported may represent nothing more than the relics of these.

Carpogonia tend to be produced in areas where active development is taking place. Such localisation is probably related to the fact that a carpogonium is formed from an apical cell and that this conversion does not seem able to occur with an inactive apical cell. Actively-dividing apical cells occur in greatest abundance in the terminal meristematic regions of thalli so that carpogonia tend to be formed most frequently in such areas. There is one important variant on this concept, the nemathecium, which as stated previously (see p. 124) consists of a superficial tissue mass formed as a result of a resurgence of activity in apical cells in certain mature areas of the thallus. In this way, meristematic tissue is formed in an area where cell division, particularly of the apical cells of the filaments of limited growth, has normally ceased. Nemathecia may be of various forms, the most common being the flat superficial mass of tissue. In many species of *Gigartina* the nemathecium occurs on a papilla which is stalked, with a narrow base, reaching 1 cm in length.

There have been many discussions of the homology of the carpogonium (see Fritsch, 1945; Kylin, 1956) with divergent views as to whether this should be interpreted as derived from one or two cells. Briefly, the protagonists of the latter view draw attention to the presence of trichogyne nuclei and the widespread occurrence of a constriction at the base of the trichogyne, whilst the opponents suggest that the spermatium may also exhibit an analogous binucleate condition and that this is of little significance.

The male organs

The male gametes, or spermatia, are spherical or ovoid 2-5 μm in diameter, formed singly within the gametangia, or spermatangia (Fig. 25). The spermatangia are produced on spermatangial mother cells which in some of the Nemaliales are not distinguishable from vegetative cells of the thallus, although in most Florideophyceae they terminate a compact system of branched filaments composed of small cells. The spermatangial mother cells may have reduced chloroplasts or even be totally colourless, whilst they and the derivative spermatangia are usually uninucleate irrespective of the nuclear constitution of the normal vegetative cells. From two to five spermatangia are cut off successively from the sub-terminal portions of the mother cell in most Florideophyceae. The formation of spermatangia in rows has been reported in several genera, such as *Endocladia* (Kylin, 1928a) where the spermatangia are cut off repetitively from the apex of the spermatangial mother-cell to produce linear rows of three to eight spermatangia. A somewhat different situation has been reported in *Rissoella verruculosa* and *Harveyella mirabilis* (Schotter, 1964) and certain members of the Corallinaceae (Adey and Johansen, 1972) where the spermatangia are formed in chains within the original cell wall of the spermatangial mother cell. The spermatangia of *Gelidium* and *Pterocladia* are in rows, each consisting of two spermatangia, but these appear to be the result of a transverse division of a cell cut off by a mother cell rather than two successive divisions of the latter. Apart from these two genera, each spermatangium is formed as a single protuberance of the mother cell into which a derivative nucleus passes before the protuberance is cut off by invagination of the wall. Young spermatangia are slightly elongate and measure 5 to 10 μm in diameter, with a relatively thick wall, often more than 1 μm thick. Spermatangia are usually colourless, although chloroplasts have often been reported in members of the Nemaliales while chloroplasts and proplastids have been recognised in young spermatangia of *Ptilota densa* (Scott and Dixon, 1972b). Because of the frequent formation of spermatangia in clusters and the aggregation of spermatangial mother cells into groups, the resulting patches of spermatangia may be sufficiently large for the colour difference to be visible to the naked eye. Conversely, when spermatangia are produced singly and the mother cells are scattered over the surface

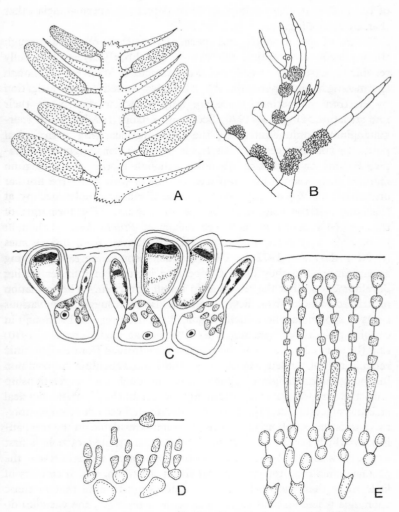

Fig. 25. *Spermatangia in Florideophyceae.* A, *spermatangia formed in dense clusters surrounding an axis*, Bonnemaisonia nootkana (× 40); B, *spermatangia formed in clusters*, Callithamnion tetragonum (× 180); C, *development of spermatangia*, Polyneura hilliae (× 1700); D, *spermatangia in pairs*, Gelidium robustum (× 800); E, *spermatangia in chains*, Endocladia muricata (× 700)

(A, D, E, *after Kylin;* B, *after Rosenvinge;* C, *after Grubb*) of the thallus it may be impossible to detect the spermatangia other than by sectioning.

Details of spermatangial and spermatial structure obtained through studies using only optical microscopy (Grubb, 1925) are not fully reliable because of the small size of the structures involved. Although spermatangia and spermatia are admirably suited for investigation by electron microscopy, there are only very few reports on their fine structure (Kugrens, 1970; Scott and Dixon, 1972b). Each spermatangium is uninucleate, with the single nucleus lying at the apical pole. In young stages, the basal pole of the spermatangium is occupied by one or more large vacuoles although as the spermatangium matures these vacuoles are replaced by others with a striated fibrillar organisation. When fully mature, these fibrillar vacuoles occupy at least one half the volume of the spermatangium. The formation of the striated vacuoles in the spermatangia of *Ptilota densa* is thought to be due to repeated fusions of vesicles derived from dictyosomes (Scott and Dixon, 1972b). However, dictyosomes are rare in young spermatangia and they do not appear to increase in number during development so that the relationship is still uncertain. The formation of striated vacuoles from dictyosomes has also been suggested in various types of sporangia where such vacuoles have been reported although in certain 'parasitic' genera it has also been suggested (Kugrens, 1970) that the striated vacuoles of spermatangia are formed from endoplasmic reticulum. Distended cisternae of endoplasmic reticulum are common in young spermatangia of *Ptilota densa* although such a relationship with the striated vacuoles could not be established. Histochemical staining suggests that, prior to release, the striated vacuoles contain polysaccharides although no further information is available on the mechanism by which release is effected. The fibrillar striated vacuole is first extruded while the spermatium is still in the spermatangium and the plasmalemma then reforms around the spermatium, which consists of little more than the nucleus. It is interesting to note that in those sporangia where striated vacuoles have been reported, the vacuoles do not appear to be released prior to the escape of a spore. Liberation of the spermatium follows the secretion of the striated vacuole and involves the rupture of the spermatangial wall. It would appear that the moderately electron-dense 'cores' of the striated vacuoles are simply cast aside during release, while the other portions of these vacuoles contribute to the mucilaginous coat of the spermatium. At about the moment of release the nuclear membrane of the spermatial nucleus disappears, which would appear to confirm old-established observations based on light microscopy and their suggestions that the nucleus was in a prophase condition after release. The nature of the bounding surface of the liberated spermatium has long been the subject of controversy, some authors claiming that some sort of wall was present even prior to release while others denied the existence of such a structure. The fine structural investigations indicate clearly that the liberated spermatia do not possess a rigid cellulosic cell wall but simply a membrane with a thick external layer of mucilage. This is of an extremely watery consistency at the time of release. There have been many descriptions of amoeboid movement of released spermatia, as well as statements that spermatia must be somewhat sticky because of the frequency with which they have been reported adhering to various thallus surfaces in the vicinity of spermatangia. The occurrence of flagellated gametes was reported in the last century for various Rhodophyta but these descriptions were probably based upon the motile reproductive bodies released by simple fungi present within the various red algae. It is very probable that the recent report of the detection of a flagellum in a cell of a spermatangial branch of *Bonnemaisonia hamifera* (Simon-Bichard-Breaud, 1971) is a consequence of a similar infection. Despite the very watery consistency of the outer mucilage at the moment of release of the spermatium, there is obviously a more rigid bounding material left after the contents of a spermatium have been transferred to a trichogyne during fertilisation. There is no knowledge of when this is produced or of its composition. In most cases, the empty 'shell' left behind is soon lost, as a consequence of the rapid decay of the trichogyne once nuclear transfer has taken place. In *Batrachospermum*, where the trichogyne is more persistent, the empty 'shells' of spermatia can remain for several weeks. It has been suggested in various genera, particularly in members of the Corallinaceae (Suneson, 1937), that the entire spermatangium is first released and the spermatium is then discharged from the spermatangium after some time. The spermatangia are tightly packed and easily dislodged in making preparations, so that such reports must be treated with caution until fine structural details confirming this are available. The spermatangia are overlaid by a layer of wall material in certain genera, often of some thickness. With the progressive release of spermatia, the thickness of this overlying layer increases. The way by which spermatia traverse this region is not known, although 'tracks' through the mucilage may be detected under suitable conditions of microscope illumination. The wall material in these thickened areas appears to be of softer consistency than normal and particularly susceptible to bacterial attack. The vague ghost-like outline of the thallus periphery and the greater depth of visible mucilage may usually be sufficient to provide an indication of the occurrence of spermatangia even with optical microscopy at quite low magnification. In some cases, secondary spermatangia may be formed by proliferation within the empty wall remaining after the spermatium has been shed. As many as seven distinct layers of empty walls have been observed in electronmicroscope preparations of *Ptilota densa* (Scott and Dixon, 1972b). Usually, this phenomenon is too difficult to observe by optical microscopy.

Although there have been so few investigations of spermatangial and spermatial fine structure, the results have been of considerable significance. On the one hand, many of the statements based on optical microscopy which were regarded with some suspicion have been confirmed while much additional information has been obtained. The presence of the striated fibrillar vacuoles and the behaviour of these in spermatial release are probably the most significant. There is still considerable scope for further investigation, both of structure as well as of the spermatial release mechanism.

Fertilisation

The wall material of the trichogyne, particularly in the terminal region, is more diffuse than normal and apparently somewhat mucilaginous. The spermatia make contact and adhere to this. There would appear to be a gradient of attraction along the trichogyne with the maximum at the distal end, so that spermatia rarely adhere to the proximal end while adhesion to the swollen base containing the carpogonial nucleus is apparently unknown. The mechanism by which the spermatia are brought into juxtaposition with the tip of the trichogyne is completely unknown. The fertilisation process in the Florideophyceae is efficient and it is difficult to explain this, on statistical grounds, considering that the female gamete is immobile and the male gamete non-motile. Spermatia can exhibit amoeboid movement but this does not appear adequate to account for the discrepancy. It has been shown that the spermatia are produced in strands of slime in *Pleonosporium* and it has been suggested (Neushul, 1972) that fertilisation is enhanced by the transfer of whole slime strands from the spermatangial to the carpogonial plant. An open connection between the cytoplasm of the trichogyne and the spermatium is established in most examples. It would appear that a delimiting wall or membrane is present in the latter by this time despite its condition at the moment of release. The nucleus of the spermatium undergoes division after contact with the trichogyne in certain of the Nemaliales, such as *Batrachospermum* and *Nemalion*, although this has not been reported in any of the higher Florideophyceae. The report of binucleate spermatia in *Lithophyllum corallinae* (Rosenvinge, 1917) is possibly an analogous situation, whilst it has been argued that the nucleus of the spermatia being in a prophase-like condition could represent the relics of such a division. The nucleus or nuclei pass from the spermatium into the tip of the trichogyne and it would appear that the spermatial cytoplasm migrates also in that an empty wall or membrane remains. Where several spermatia make contact with a single trichogyne it would appear very probable that the nuclei and cytoplasm of each spermatium move into the trichogyne in that all the spermatia are eventually empty. The fate of supernumary nuclei is not known.

The nucleus of the carpogonium is always in an interphase condition whilst the one spermatial nucleus which is going to fuse with it is usually in an advanced state of prophase as it migrates down the trichogyne. The spermatial nucleus migrating down the trichogyne has been described as being in an interphase condition in both *Batrachospermum* and *Nemalion*. At the moment of nuclear fusion, the chromosomes derived from the spermatial nucleus are usually clearly differentiated, although too tightly aggregated to permit any estimate of number. The carpogonial nucleus begins to differentiate and the zygote nucleus which is formed goes into division almost immediately. There has been considerable discussion as to the nature of the first division of the zygote nucleus in many of the Florideophyceae; this is discussed in detail later (see p. 193).

Following fertilisation, the trichogyne is separated from the basal part of the carpogonium, either by a membrane, or by a wall, or by the development of a 'plug' across its aperture. Before this separation occurs the cytoplasm of the trichogyne is usually retracted completely, although some traces may remain in certain cases. Once separation from the basal part of the carpogonium has occurred, the trichogyne disintegrates rapidly, except in a few Florideophyceae such as *Batracho-spermum* or *Nemalion* where it is remarkably persistent.

It has been argued that fertilisation may not necessarily precede further development in many Florideophyceae and that parthenogenesis may be occurring. The evidence on which this argument is based is far from satisfactory. The failure to detect spermatangial plants is often interpreted as evidence of parthogenesis. As has been indicated previously (p. 140), large clusters of spermatangia form colourless patches visible to the naked eye, although when formed singly the spermatangia may be detected only by careful scrunity of sections and even then are often overlooked. Similarly, the development of a carposporophyte from carpogonia where there are no traces of spermatia on the trichogynes has been advanced as an argument for parthenogenesis (Rosenvinge, 1917), although the ephemeral nature of both trichogyne and the remains of attached spermatia indicates that this evidence is not sufficient. There are in addition various signs derived from both cytological studies and cultural investigations (Dixon, unpublished observations) which are best explained by reference to parthenogenesis, but the evidence is still inadequate for any positive statement of its occurrence.

THE CARPOSPOROPHYTE AND ITS DEVELOPMENT

As stated previously, the Florideophyceae are characterised by the retention of the zygote on the female gametophyte where it gives rise to a complex post-fertilisation development known as the carposporophyte. Interpretation of this structure has undergone significant changes during the past two centuries and there are still a number of uncertainties remaining. It was appreciated at a very early date by Stackhouse (1801) and Turner (1802) that in those algae now referred to as the Florideophyceae the spores, or 'seeds' as they were then called, were produced in two very different ways. In one, now described as the tetrasporangia, the spores were formed in groups of four, whereas in the other the spores were produced in a mass which may or may not possess a surrounding envelope. Many different terms were applied to the variants of the latter; the following are the terms and definitions used by Harvey:

favella: a 'berry-like' mass of spores, with no envelope, such as occurs in *Ceramium* or *Callithamnion*;

favellidium: a 'berry-like' mass of spores immersed in the thallus, as in *Naccaria*;

coccidium: a globose body, frequently half-immersed, as in *Plocamium*;

ceramidium: an urn-shaped structure containing spores.

This terminology has now disappeared, although the term 'cystocarp' used first by the younger Agardh in 1844 to describe spore masses surrounded by a definite envelope, is still retained by some investigators. Following the discovery of the sexual organs and the process of fertilisation by Bornet and Thuret it became clear that this structure was formed as a consequence of gametic fusion. The details of the development in selected genera were first elucidated by Schmitz (1883) although his interpretations were not entirely correct. The most important discoveries made by Schmitz were:

(*a*) that when an envelope surrounds the spore mass, only the spore mass is formed from the zygote, whilst the envelope is produced by an outgrowth of the female gametophyte;

(*b*) that the zygote may develop directly into the spore mass or indirectly through a complex process involving further cell fusions.

Schmitz incorrectly assumed that the latter were equivalent to the preceding gametic fusion. The correct interpretation was provided somewhat later by Oltmanns (1898, 1904) who showed that these secondary fusions involved nuclear transfer but not nuclear fusion. Oltmanns also regarded this post-fertilisation structure as equivalent to a plant producing tetrasporangia or gametangia in that he called it the sporophyte, and referred to the tetrasporangiate plants as '*Nebenfruchtform*'. The word 'carposporophyte' was coined by Janet (1914) and both it and the concept of this being a phase or generation equivalent to the gametangial and tetrasporangial plants have been accepted almost universally. The cell mass which constitutes the carposporophyte is formed from one or more primordia. The term gonimoblast was used first by Schmitz (1883) to describe this tissue as a whole. Subsequent usage has varied somewhat, in that Kylin (1937a, 1956) and Drew (1954a) follow the Schmitz concept, whereas Fritsch (1945) and Smith (1944) refer to the outgrowth of each primordium as a 'gonimoblast'. It would appear preferable to accept the former interpretation.

The carposporophyte terminates with the production of carposporangia. The carposporangia may be formed only from the apical cells of the carposporophyte although, in some genera, intercalary cells may be converted in addition to the latter. In most genera, each carposporangium liberates a single carpospore although in a few examples four such spores are produced from each carposporangium. The liberation of carpospores takes place through the rupture of the sporangial wall. The development of a new carposporangium within

F

the old sporangial wall, by proliferation through the pit connection from the cell below, has been reported in some genera but by comparison with other types of sporangia such regeneration occurs only very rarely.   In the early stages of development of the carposporophyte it is often very difficult to predict the ultimate fate of the tissue which is being produced.   A mass of cells, organised into filaments, with dense cytoplasmic contents is formed; some of these cells will develop into carposporangia, others will not.   Even in those cases where the greater part of the carposporophyte is converted into carposporangia (cf. *Lemanea*) there will always be a few cells at the base which remain unconverted.

It is understandable that with the changes in interpretation which have occurred, the nomenclature of various structures should be confused.   In addition, many difficulties have resulted from the microscopic size of the structures and the amazing complexity of some of the developmental processes.   Because of such difficulties, the accumulation of data has been extremely slow, so that even today there are some genera and families for which there is no information available and many for which the data are either fragmentary or even contradictory.   The following account of the morphological development of the carposporophyte in selected genera will give only an outline of the principal types which are found.

The simplest type of carposporophyte development occurs in species of *Acrochaetium* and *Batrachospermum* (Fig. 26).   The development is most clear in *Batrachospermum* where the basal part of the carpogonium gives rise to one or more lateral protuberances, into which pass nuclei derived from the zygote nucleus.   There have been various statements as to the nature of this particular nuclear division but consideration of this question will be left until later (p. 193), where it will be possible to give a comprehensive survey of the situation with regard to nuclear divisions in the carpogonium throughout the Florideophyceae.   The protuberances are then cut off and function as apical cells.   Each develops to produce the gonimoblast, which in this genus consists of a system of branched filaments.   These form a compact mass, except in one species, *B. orthostichum*, where the gonimoblast is diffuse and spreading.   The carposporangia are formed by the conversion of apical cells, either in a terminal or lateral position and each produces a single carpospore.   In addition to the development which takes place from the carpogonium, cells below it also begin to divide following the occurrence of fertilisation.   The one or two cells immediately below

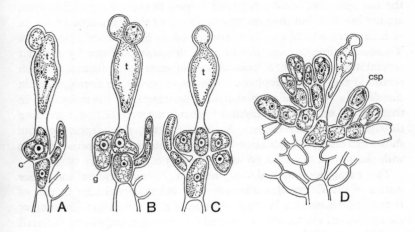

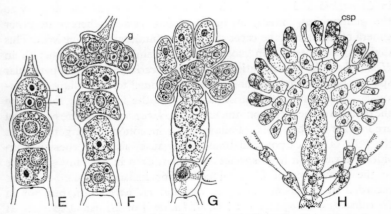

Fig. 26. *Carposporophyte development in* Batrachospermum *and* Nemalion. A–D, Batrachospermum moniliforme; (A–C, ×1000; D, ×800); E–H, Nemalion helminthoides; (E, F, ×825, G, ×700; H, ×450). (*ct, carpogonium; csp, carpospore; u, upper cell; l, lower cell; g, primary gonimoblast; t, trichogyne*)

(A–H, *after Kylin*)

the carpogonium, usually regarded as part of the carpogonial branch, are not involved but the vegetative cells below these produce a number of lateral protuberances which are cut off to form lateral apical cells. These sometimes remain in this state although they usually segment several times to form a loose cluster of enveloping filaments which surround the carposporophyte. The carposporophyte development in *Sirodotia* is very similar to that of *Batrachospermum orthostichum* in that the one or more gonimoblast filaments are very diffuse, spreading through the cortical tissue. As in the latter, the carposporophyte shows increasing specialisation being differentiated into horizontal axes with short erect filaments on which the carposporangia are borne.

The carposporophyte development in *Helminthocladia* and other genera of the Helminthocladiaceae (Fig. 26) is very similar to that of *Batrachospermum*. As in the discussion of that genus, consideration of the cytological events which occur in the carpogonium will be deferred until later (p. 193). In these genera of the Helminthocladiaceae there are a number of variants from the development of the carposporophyte of *Batrachospermum* discussed previously. The fertilised carpogonium in *Cumagloia* and probably also in *Dermonema* gives rise directly to the gonimoblast initials, as in *Batrachospermum*, whereas in other genera of the family the carpogonium first undergoes a division. This division is transverse in most of the remaining genera although in *Helminthocladia* its orientation is somewhat variable, even in a particular species, ranging from the obliquely longitudinal to the obliquely transverse. The gonimoblast arises only from the upper cell formed by the division of the carpogonium, except in *H. papenfussii* where primordia are formed on both cells. Following the inception of the gonimoblast the cells of the carpogonial branch may fuse, although there is considerable variation from species to species, even in the same genus, as to the occurrence of a fusion cell. For instance, fusion cells are recorded in *Helminthocladia australis* and *H. papenfussii*, but not in *H. calvadosii*, *H. agardhiana* [= *H. hudsoni*] or *H. macrocephala*. It appears that in most genera these fusion cells are initiated by the enlargement of the pit connections so that the cells which fuse are already in contact. The extent to which the enlargement of pit connections takes place can be extremely variable. The degree of enlargement in *Helminthocladia papenfussii* is so great that a large conical cell is formed whereas in other species the original cell outline is retained. In those taxa where a fusion cell develops, this is formed from the lower half of the carpogonium and the two hypogynous cells, although in *Trichogloea requienii* and *Helminthocladia papenfussii* the upper half of the carpogonium is involved also. The situation with regard to fusion cells in *Dermonema* is a little confused. According to Svedelius (1939) the undivided carpogonium gives rise directly to the gonimoblast initials in most examples. The carpogonial branch is markedly recurved and it was reported that in a few instances the carpogonium fused with either the supporting cell or the first cell of the carpogonial branch before initiation of the gonimoblast. Subsequently, Desikachary (1962) has stated that he observed not a single instance of such fusions but that widening of the pit connections between the cells of the carpogonial branch occurred. The latter resemble closely the widening which precedes the formation of the fusion cells in other genera of the Helminthocladiaceae. Following the initiation of gonimoblast filaments, enveloping involucral filaments may or may not be formed, whilst in *Nemalion multifidum*, *Trichogloea requienii* and *T. herveyi* it has been shown that these may develop prior to fertilisation. Enveloping filaments are absent in *Dermonema* but recorded in some species of all the remaining genera. The extent to which their development takes place may vary, as in *Nemalion*, whilst in *Helminthocladia calvadosii* they have been reported by some investigators and their presence denied by others. The enveloping filaments may be formed from cells above the support cell, or from cells both above and below the latter, or even from cells of the carpogonial branch. The carposporophyte is compact and dense in all genera except *Dermonema* and *Cumagloia* where elongate gonimoblasts extend horizontally through the thallus, in a manner similar to that found in *Sirodotia*. In most species of the Helminthocladiaceae, each carposporangium gives rise to a single carpospore although in *Helminthocladia agardhiana* [= *H. hudsoni*], *Liagora tetrasporifera* and several other species of the latter genus the carposporangia are quadripartite, each giving rise to four spores.

In the examples which have been considered, the carposporophyte arises either from the entire carpogonium or from a portion of it. With the exception of the disputed reports in *Dermonema*, such fusions as occur do not precede the initiation of the carposporophyte but occur after or simultaneously with its development. The nucleus resulting from fertilisation, or any of its derivatives, is not transferred to any other cell of the gametophyte. In addition, there appear to be no specialised storage tissues whose contents disappear during the development of the carposporophyte, presumably being utilised as a source of nutrition for the latter.

Such 'nutritive tissue' does occur in the Naccariaceae, Bonne-maisoniaceae, Chaetangiaceae and Gelidiaceae, which in this and other respects exhibit various features of a more complex nature not met with in the Helminthocladiaceae. Throughout the Naccariaceae and Bonnemaisoniaceae, short lateral branches composed of cubical or spherical cells with dense contents are borne on the hypogynous cell, or cells, of the carpogonial branch. The details of development of the carposporophyte vary from species to species, that of *Asparagopsis* being very much more complex than that of *Atractophora*. The carpogonium in the latter fuses with the supporting cell before giving rise to the two or three gonimoblast initials, whilst in *Naccaria* the carpogonium fuses with the hypogynous cell following enlargement of the pit connection between the two. The pit connection between the hypogynous cell and the cells of the 'nutritive tissue' may also widen, although the cellular arrangement is not changed. In *Bonnemaisonia*, on the other hand, complete fusion of these cells occurs. The cells of the 'nutritive tissue' enlarge and increase in number subsequent to fertilisation and, during the early segmentation of the carposporophyte, fuse with one another and with the hypogynous cell. The gonimoblast initials segment to form several cell layers and the lowest also fuses with the 'nutritive tissue' to produce a large fusion cell of irregular shape. The uppermost cells of the gonimoblast give rise to a cluster of filaments, the carposporangia being formed from their apical cells. The zygote nucleus in *Asparagopsis* actually migrates into the hypogynous cell from the carpogonium which then degenerates, after which a large fusion cell similar in structure and origin to that of *Bonnemaisonia* is formed. The flask-shaped structure which envelops the carposporophyte in both families is very much more substantial than any of the enveloping structures found in the Helminthocladiaceae. Even before fertilisation, a considerable amount of tissue is produced and, once the development of the carposporophyte has commenced, growth of the envelope is rapid. This produces a very characteristic flask-shaped envelope, several cell layers in thickness, with a terminal aperture through which the spores escape.

The Gelidiaceae resembles the Naccariaceae in a number of respects, both with regard to the presence of nutritive tissue, which develops after fertilisation from a certain cell below each carpogonium as a dense recurved 'umbrella-like' mass of cells in short much-branched filaments and to the development of the carposporophyte. The initial account of carposporophyte development in *Gelidium*, given by Kylin

(1928a) has been shown to be incorrect in a number of respects. The gonimoblast is *not* formed directly from the unchanged carpogonium as Kylin suggested, but a large multinucleate cell is formed and the gonimoblast is produced from non-septate processes issuing from this. According to Fan (1961), the formation of the multinucleate cell always involves a widening of the pit connection between the carpogonium and the cell on which it is borne and their ultimate fusion, with other fusions occurring infrequently, but in British material (Dixon, 1959) fusion with the supporting cell was by no means a constant feature. Fusions with adjacent cells were observed in the British material but no evidence was obtained to indicate the occurrence of direct fusion of the gomino-blast with the nutritive tissue such as was figured by Fan. The essential differences between the carposporophytes of the Naccariaceae and the Gelidiaceae are that in the former they are produced singly, at the tips of axes, with a specialised envelope, whereas in the Gelidiaceae several carposporophytes may develop internally to produce a compound structure, with no additional adventitious envelope. Following the development of one or more carposporophytes the original cortex of the axis, consisting of from five to ten cell layers, is pushed up. In all genera of the Gelidiaceae except *Pterocladia* this lifting occurs on both sides of the axis to form a pair of cavities separated by the axial plate and carposporophytes. Spores are discharged from the carpo-sporangia into these cavities and then released through two ostioles which develop on either side of the axis. The inflation in *Pterocladia* is much more one-sided, with only a single cavity on one side of the axis in most species although a second, much smaller cavity may develop on the opposite side to the first in *P. lindaueri*, but there appears to be no trace of an ostiole associated with the smaller cavity in this species. One further difference between *Pterocladia* and the other genera of the Gelidiaceae is that in this genus more than one ostiole may develop in association with a cavity and in extreme cases a row of three or four such ostioles may be formed on the side of the axis on which the latter develops.

The hypogynous cell in *Scinaia* and other genera of the Chaetangiaceae gives rise, not to filaments of cells with dense contents such as occur in the Bonnemaisoniaceae, Naccariaceae and Gelidiaceae, but simply to three cells with dense contents, which often contain several nuclei. The exact name to be applied to these will be discussed later (p. 238). After fertilisation, the zygote nucleus migrates from the carpogonium through the enlarged pit connection into the hypogynous cell where it has been claimed that it undergoes a meiotic division. As in the previous discussion of examples the type of nuclear division which occurs has been omitted for detailed discussion later (p. 193). The gonimoblast initial is formed from the hypogynous cell and grows back into the old carpogonium through the enlarged pit connection. With further segmentation, the gonimoblast escapes laterally through the side wall of the carpogonium whilst the pit connections between the hypogynous cell and its three derivative cells enlarge to produce a partial fusion. The branches of the gonimoblast form carposporangia both from apical as well as sub-terminal cells so that these are produced in short chains of two to four sporangia. The envelope surrounding the carposporophyte is made up of a dense mass of filaments which arise entirely from the first cell of the carpogonial branch, but unlike many of the previous examples in which such an envelope occurs this is embedded completely in the parent thallus in the genera of the Chaetangiaceae.

The preceding discussion has been concerned with genera and families attributed to the order Nemaliales in the broadest sense (see p. 239). One obvious conclusion from this discussion is that throughout each family there is considerable uniformity in relation to carposporophyte development, with increasing specialisation in terms of the specificity of cellular fusions.

Turning now to other orders, fusion of the gonimoblast with specified cells of the gametophyte at some distance from the carpogonium occurs in a number of genera. Feldmann (1952a) has shown in *Bertholdia* (Gigartinales) that four gonimoblasts are produced after the fusion of the fertilised carpogonium with cells of the carpogonial branch. These gonimoblasts spread out laterally through the thallus and fuse with certain intercalary cells of the female gametophyte, although there is no nuclear transfer to the cells with which the gonimoblast fuses. The clusters of carposporangia are formed from short radially-directed branches of the gonimoblast but the sites of origin of these bear no relationship to the positions of the cells which fuse with the gonimoblast. Drew (1954a) has indicated that a number of examples, similar to *Bertholdia* in their carposporophyte development, demonstrate a gradual shift in the position at which development of the carposporangia is initiated, from the gonimoblast itself to the cell with which this fuses. The carposporangium-producing branches in *Polyides* (Gigartinales) (Fig. 27, c) and *Cruoria* (Gigartinales) are, as in *Bertholdia*, some distance from the cells with which the gonimoblast fuses, whilst in *Nemastoma*

(Gigartinales) they develop relatively close to these cells. Finally, in *Acrosymphyton* (Cryptonemiales) carposporangium production is from the gonimoblast but at the point of fusion with the cell with which it fuses. This state is little removed from the condition found in many of the Cryptonemiales and Gigartinales where the cell which fuses with the gonimoblast acts as the site for the origin of the carposporangia, the fusion involving, of course, the transfer to that cell of the zygote nucleus or a derivative of the latter. The classic example of this type of carposporophyte development is exemplified by *Dudresnaya verticillata* (= *D. coccinea*, Cryptonemiales), one of the species first investigated by Oltmanns. The cells to which the zygote nucleus, or its derivative, is transferred has been termed the 'auxiliary cell' (Fig. 27, D, E). The application of that term has been the subject of much controversy, and this will be discussed in detail later (see p. 160). These auxiliary cells in *Dudresnaya* occur on short filaments which develop from the basal cells of the vegetative laterals in a position analogous to that of the carpogonial branch. The branches bearing auxiliary cells are always more numerous than those with carpogonia. Following fertilisation, the carpogonium produces a short single-celled gonimoblast.

Considering first those members of the Gigartinales in which procarps occur, these represent a whole series of variants among which are some of the most complex carposporophytes in the Florideophyceae. One of the simplest examples is exhibited by *Cystoclonium*, in which the auxiliary cell is formed from the lowermost cell of a vegetative filament which is associated with the supporting cell of the carpogonial branch. In other genera of this order, such as those of the Gigartinaceae and Phyllophoraceae, the auxiliary cell is formed from the supporting cell on which the carpogonial branch is borne. In either case, the carpogonium and auxiliary cell lie in close proximity to one another although the means by which contact between the two is effected is not known. A gonimoblast initial is cut off eventually from the auxiliary cell in *Cystoclonium* developing into a branched filament of cells which ramifies through the adjacent vegetative regions of the thallus. Groups of carposporangia are formed from the tips of this branched system, the cells of which fuse with one another and with the auxiliary cell to produce an irregularly branched fusion cell. The tissues of the female gametophyte lying over the procarp undergo rapid cell division so that the mature carposporophyte appears as a prominent swelling devoid of an aperture on the side of the thallus (Fig. 28, B). The developing carposporophyte is often surrounded by abundant hyphae,

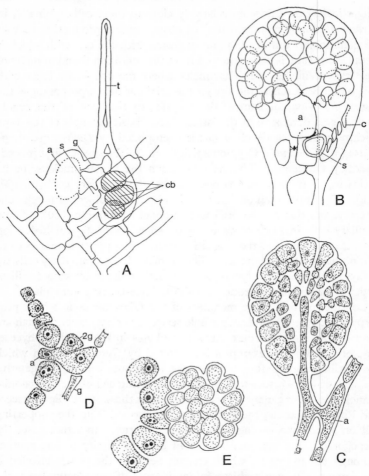

Fig. 27. *Carposporophyte development involving an auxiliary cell.*
A, Antithamnion plumula, *showing connection between carpogonium and
auxiliary cell* (× 700); B, Ptilothamnionopsis lejolisea, *showing origin of
carposporangial mass from the auxiliary cell and the remains of the
carpogonial branch* (× 60); C, Polyides rotundus, *showing fusion of
primary gonimoblast with auxiliary cell and the derivation of a carposporangial mass from the continued growth of that gonimoblast* (× 285); D,
Dudresnaya verticillata, *showing fusion of primary gonimoblast with
auxiliary cell and both the continued growth of the primary and origin of
secondary gonimoblast* (× 600); E, Dudresnaya verticillata, *showing
origin of carposporangial mass from the secondary gonimoblast arising
from the auxiliary cell* (× 600). (a, *auxiliary cell;* c, *carpogonium;* cb,
carpogonial branch; s, *support cell;* g, *primary gonimoblast;* 2g, *secondary
gonimoblast;* t, *trichogyne*)
(A, *after Phillips;* B, *after Dixon;* C–E, *after Kylin*)

although in *Cystoclonium* there appear to be no fusions between gonimo-blast and hyphae. This is not the case in some of the more complex carposporophytes, such as occur in *Rhodophyllis* or *Calliblepharis*, where some branches of the gonimoblast do not give rise to carposporangia but become attached by secondary pit connections to the surrounding sterile tissue of the female gametophyte. In the most elaborate types, such as *Gracilaria foliifera*, the carposporophyte is attached to the surrounding sterile tissue not only by attachments originating in the carposporophyte and growing outwards, but also by other filaments growing inwardly from the sterile tissue of the surrounding envelope.

In all members of the Gigartinales which have been considered, carposporangia, each of which liberates a single carpospore, are formed from the apical cells of the gonimoblast, as well as from intercalary cells in many instances. This is also the case with some members of the Phyllophoraceae, such as *Stenogramme interrupta* and *Phyllophora membranifolia*, although the development in certain species of the latter and in other genera does exhibit significant deviation from the general situation found in the procarpic Gigartinales. The procarp of *Phyllophora truncata* is similar in structure to that of *P. membranifolia*. There is some disagreement as to the details of subsequent develop-ment, although it is agreed that the gonimoblast filaments spread through the female gametophyte and establish pit connections with its cells. The overlying cortex is raised to produce small cushions of cells which fuse to form large globular nemathecia. The surface of each nemathecium is made up of elongate, radially directed filaments of cells, of which all except the outermost three or four are converted, not into normal carposporangia each producing a single carpospore, but into tetrasporangia. *Phyllophora truncata*, in this respect, is analogous to *Liagora tetrasporifera* and other members of the Helminthocladiaceae in which quadripartite carposporangia occur. The situation in *Gymnogongrus* is similar to that in *Phyllophora*, carposporophytes typical of the procarpic Gigartinales occurring in some species, such as *Gymnogongrus linearis*, *G. pusillus* and *G. norvegicus*, but not others. The procarps of *G. griffithsiae*, *G. platyphyllus* and *G. devoniensis* are rudi-mentary, spermatangia are unknown and there is no evidence for fertilisation except in an illustration of *G. devoniensis* given by Schotter. The supporting cell of the carpogonial branch gives rise to filaments, which are equivalent to the gonimoblasts of *Phyllophora truncata*, and which aggregate to form a nemathecium at the surface of the thallus. The nemathecium is made up of radially-directed filaments of cells which form tetrasporangia or monosporangia. This reduction in pro-carp structure and carposporophyte development is even more extreme in *Ahnfeltia plicata*. Here there is no recognisable procarp, as such, the nemathecia being formed by proliferation from apical cells of the filaments of limited growth of the thallus which in particular areas have cell contents more dense than those of their surrounding apical cells. A nemathecium is formed which produces monosporangia super-ficially, the report of tetrasporangia being erroneous. The intervening steps in the development of a nemathecium in *Ahnfeltia* are very confused (Rosenvinge, 1931; Schotter, 1968) and it is not possible to give an adequate explanation on the basis of existing information. The nemathecia of these species of the Phyllophoraceae are sufficiently different from the plant upon which they are borne that for some time they were regarded as independent parasitic Florideophyceae, and binomials assigned to them.

There are two distinct types of procarpic arrangement in the Crypto-nemiales. That found in the Gloiosiphoniaceae, Endocladiaceae and Kallymeniaceae is very similar to those examples described previously in the Gigartinales, while a very different type of arrangement and subsequent carposporophyte development is shown by genera of the Corallinaceae. This family exhibits specialisation in reproduction similar to that found with their vegetative structure. Large numbers of procarps are initiated within each female conceptacle, although only those in the centre appear to reach maturity. There are various different arrangements of cells in the procarp in this family although it has been shown by Suneson (1937) that these can be interpreted as a reduction series from a basic type of construction. This basic con-struction consists of a supporting cell bearing three branches, not all of which are represented in every case. The most complex type is found in various species of *Melobesia*, where two of the three branches are two-celled, the apical cell forming a carpogonium, whilst the third consists of a single-celled elongate primordium which undergoes no further development. Each procarp in *Corallina* contains only one two-celled carpogonial branch and the central single-celled primordium, the other carpogonial branch being represented only by a single-celled primordium or suppressed completely. An even greater reduction is shown in *Choreonema* where a single two-celled carpogonial branch is present with the supporting cell producing a long protuberance which is equivalent to the single-celled central primordium in the previous examples but which, in this instance, is not cut off from the supporting

cell. The final condition is met with in various species of *Lithothamnium* and *Melobesia* where even this protuberance is lacking and
the supporting cell appears simply as the basal cell of a three-celled
structure. Fertilisation of a single carpogonium among the many
present in each female conceptacle appears to be sufficient for the
initiation of further development although, throughout the family,
investigation of carposporophyte development is extremely difficult and
there are many incomplete accounts and unanswered questions, particularly with the early stages. In one example (*Melobesia lejolisii*) it
has been demonstrated that the carpogonium produces a short outgrowth which fuses with its auxiliary cell (Suneson, 1937) whilst in
Choreonema the carpogonium is said to fuse directly with the auxiliary
cell lying alongside although the nucleus does not migrate but remains
in the carpogonium. Subsequently, fusions occur laterally between
the various auxiliary cells although there is some argument as to the
position at which the fusion cell develops. In *Melobesia* it appears to
be based around the primary auxiliary cell, with the remains of mature
and immature procarps on its upper face, whilst in *Choreonema* it is
said to be formed more by the enlargement of the carpogonium, short
processes put out by the latter making contact with surrounding
auxiliary cells which eventually degenerate. The fusion cell, which-
ever way it may be formed, then gives rise to carposporangia on the
flanks. Short filaments of cells are produced first and the apical cell
then converts into a carposporangium followed by the remaining cells
in basipetal sequence. The carpogonium and that cell then fuse with
the two hypogynous cells in the carpogonial branch immediately below
the carpogonium. These two hypogynous cells have relatively dense
contents. Following the fusion the original cell outlines remain reasonably distinct, although the cytoplasm tends to accumulate in the old
carpogonium and the gonimoblast cell. Elongate gonimoblast filaments then develop from the latter position and spread through the
thallus, making contact with a number of auxiliary cells at a distance
from the carpogonial branch. The auxiliary cell puts out a short protuberance which fuses with the gonimoblast, receiving from it a derivative of the zygote nucleus. One interesting aspect of *Dudresnaya* is
that the elongate gonimoblast filaments make contact with a number of
auxiliary cells in turn (Fig. 27, D, E). Each auxiliary cell then produces
two or three primordia which segment to form a short branched system
of tissue from which the carposporangia are produced. Carposporophyte development in *Dumontia* (Cryptonemiales) is very similar to

that of *Dudresnaya*, except that the carpogonium fuses not with the immediate hypogynous cells but with the cell of the carpogonial branch three cells distant from itself, although other cells may also be involved. In addition, the elongate gonimoblasts do not make contact in turn with a number of auxiliary cells to the same extent as in *Dudresnaya*. In *Dudresnaya*, *Dumontia* and other members of the Dumontiaceae the branches bearing carpogonia and auxiliary cells are homologous. Furthermore, in this family the cells which fuse with the carpogonium or the initial gonimoblast are clearly equivalent to auxiliary cells. In *Grateloupia* (Cryptonemiales) and other members of the Cryptonemiaceae such initial fusions are absent and it would appear that in this family complete separation of the carpogonium and auxiliary cell functions has occurred.

The auxiliary cells in *Dudresnaya* and *Dumontia* develop in relation to special accessory filaments which have been shown to be equivalent to carpogonial branches. The carposporophyte development in *Platoma* (Gigartinales) is almost identical with these two examples except that the auxiliary cell is part of an ordinary vegetative filament of the thallus. The difference between the disposition of the auxiliary cells in *Dumontia* and *Platoma* would appear to be of a trivial nature, although this feature has been proposed, by Kylin, as the fundamental criterion to distinguish between the Gigartinales and the Cryptonemiales.

The most specialised carposporophytes in which the auxiliary cells are some distance from the carpogonium occur in the families Solieriaceae and Rhabdoniaceae of the Gigartinales. The tissue of the female gametophyte around each auxiliary cell undergoes correlated growth following fusion of that cell with the gonimoblast. In this way a flask-shaped envelope with a definite ostiole is formed around each auxiliary cell, equivalent to the envelopes which develop around whole carposporophytes in other instances.

The final type of carposporophyte development to be considered is that in which the zygote nucleus, or a derivative of this, is transferred to a cell adjacent to the carpogonium. There is a large number of different variants of this, occurring in a wide range of Florideophyceae. Strictly speaking, the situation in *Asparagopsis* or in members of the Chaetangiaceae which have been discussed previously should be considered under this heading. However, the close relationship between, say, *Bonnemaisonia* where the hypogynous cell fuses with the carpogonium but does not receive the zygote nucleus and *Asparagopsis* where it does suggests that the arrangement adopted here is justifiable. The relationship between these two examples indicates one reason why definition of the term 'auxiliary cell' poses so many problems. Where the carpogonium and the auxiliary cell form part of the same branch system, the aggregate is frequently described as a 'procarp'. Procarpic arrangements are found in many of the Gigartinales and Cryptonemiales, and in all genera of the Rhodymeniales and Ceramiales.

The structure of the procarp and the development of the carposporophyte are extremely uniform throughout the Rhodymeniales. The supporting cell is formed by the transformation of a cell of the inner cortex which enlarges considerably, develops contents denser than those of its neighbours and becomes obviously multinucleate. The supporting cell then forms one or two primordia which may or may not go on to produce vegetative laterals before giving rise to a single carpogonial branch which may be either three-celled or four-celled. The one or two primordia are, in fact, the mother-cells of the auxiliary cells and although the latter are produced from the mother-cells prior to fertilisation they remain small, only enlarging and developing dense contents after that event. The zygote nucleus divides within the carpogonium and one of the products is transferred to the auxiliary cell. Fusion then occurs between the cells of the carpogonial branch to produce a prominent fusion cell whilst the auxiliary cell continues to enlarge, developing a very prominent central globule of proteinaceous material. The auxiliary cell gives rise to the gonimoblast which is much branched, although in *Chylocladia* and *Gastroclonium* a second fusion cell is formed from the innermost cells of the gonimoblast, the auxiliary cells, the mother-cells of the latter and frequently some adjacent cells of the female gametophyte. Most of the cells of the gonimoblast give rise to carposporangia in all genera of the Rhodymeniaceae and in *Lomentaria* (Champiaceae) whereas in the other genera of the latter family the carposporangia are formed only from the terminal cell. Immediately after fertilisation the overlying and surrounding cells of the cortex begin to divide to produce the flask-shaped structure surrounding the carposporophyte. In all genera of the Rhodymeniales except *Chylocladia* (Fig. 28, c) and *Gastroclonium* this is provided with an ostiole.

The procarp structure and carposporophyte development in the final order of the Florideophyceae to be considered, the Ceramiales, are very characteristic and relatively uniform throughout the whole order. The supporting cell is a pericentral cell or its equivalent, whilst the carpogonial branch is always composed of four cells, and devoid of lateral branches. There is usually only a single carpogonial branch in each

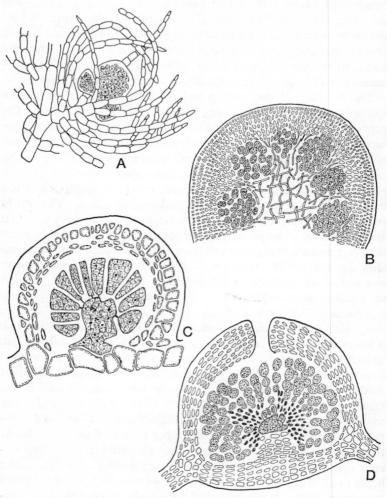

Fig. 28. *Carposporophytes of Florideophyceae.* A, Antithamnion plumula, *with carposporophyte subtended by loose adventitious filaments* (×70); B, Furcellaria fastigiata, *with carposporophytes embedded in the axis* (×30); C, Chylocladia verticillata, *with carposporophyte enclosed within pericarp* (×150); D, Phycodrys rubens, *with carposporophyte enclosed within pericarp* (×60).

(A–D, *after Kylin*)

procarp although there may be one or two sterile lateral branches arising from the supporting cell. A second carpogonial branch, less well developed than the first, may replace one of the sterile laterals in certain genera of the Delesseriaceae such as *Polyneura, Hemineura,* etc. The auxiliary cell is formed after fertilisation has taken place by the division of the supporting cell. This is one of the most characteristic features of the procarps in this order although in certain Ceramiaceae it has been shown that this division may fail to occur in which case the supporting cell itself functions directly as the auxiliary cell. In some Ceramiaceae, two auxiliary cells may be produced in each procarp. Transfer of the zygote nucleus to the auxiliary cell is said to be established by the fusion with it of a small cell cut off from the carpogonium although it would appear more probable that in many instances a cell is not cut off from the carpogonium but simply that a protuberance develops from it (Fig. 27, A). This fuses for a brief period with the auxiliary cell during which time the zygote nucleus is transferred. The auxiliary cell then produces one or more primordia from which the gonimoblast develops (Fig. 27, B). The extent to which further fusions occur varies considerably from family to family. There are no indications of any fusions whatsoever in the Ceramiaceae and Dasyaceae, except for the reports of the refusion of the support cell and auxiliary cell in various species of *Antithamnion* and related genera. Fusions of greater complexity occur in Rhodomelaceae and Dasyaceae, which frequently involve the axial cell associated with the procarp and cells of the gonimoblast in addition to the auxiliary cell and the supporting cell. The size of the carposporophyte throughout the Ceramiales varies enormously and this appears to bear some relation to the extent to which carposporangia are formed. Most cells of the carposporophyte are converted to carposporangia in the larger examples, whilst only the apical cells develop into carposporangia in the smaller. The disposition of the carposporophyte and the extent to which protective tissue develops from the female gametophyte vary from family to family. The thalli of the Ceramiaceae are obviously filamentous in construction and the carposporophyte is usually exserted and almost naked, being surrounded only by a few enveloping filaments (Fig. 28, A). The thalli of the Rhodomelaceae and Dasyaceae are most compact and the carposporophytes are enclosed by flask-shaped structures, two or even three cell layers in thickness, with distinct ostioles. The Delesseriaceae have thalli which are flat and membranaceous and the carposporophytes develop on one side of this cell plate. They are protected by a hemispherical cover with a central ostiole although, as they mature, the tissue beneath is often pushed out to form a more biconvex structure (Fig. 28, D). Thus throughout the order there are certain standard characteristics, the disposition of the carpogonial branch, the method of formation of the auxiliary cell and the carposporophyte development, with relatively minor modifications of other features in the four families.

REPRODUCTION BY GEMMAE

One of the peculiar features of the Florideophyceae is that reproduction by specialised gemmae is so very rare. Vegetative propagation involving the detachment of unspecialised portions of a thallus and their continued growth, with or without secondary attachment, is particularly common (see p. 212). However, this facility is not matched by the production and release of specialised multicellular structures for vegetative propagation. Such bodies do occur, in a species of *Melobesia* (Solms-Laubach, 1881) and in the freshwater *Hildenbrandia rivularis* (Nichols, 1965). The carposporophyte of *Batrachospermum breutelii* gives rise to large oval or spindle shaped bodies which are divided up into from three to six cells and which Skuja (1933) interprets as gemmae. Propagules said to be similar to those of *Sphacelaria* have been reported for *Polysiphonia furcellata* (Bornet, 1892) but there is no confirmation for this report. With these exceptions, organised gemmae are unknown in the Florideophyceae.

Reproduction in Bangiophyceae

Knowledge of reproductive processes and, hence, of life histories in the Bangiophyceae is extremely slight. The extreme morphological simplicity and, in many cases, the low levels of differentiation between vegetative cells and reproductive structures demand highly critical observations which, unfortunately, have not always been achieved. The situation has been further confused by the application of a great number of diverse, often ill-defined, terms, some of which have been derived from the Florideophyceae, some from other major algal groups, and some newly concocted. The present review will attempt to present, as briefly as possible, the information which is sufficiently clear to justify acceptance and to indicate the many widely accepted statements for which the evidence is not considered satisfactory at the present time. It is symptomatic of the situation with respect to reproduction in the Bangiophyceae that the most critical outstanding questions relate to the resolution of two functions which are basic and fundamental to any study of reproduction and life history.   The first question relates to the definition of a spore while the second relates to the question of gametic cell fusion in the Bangiophyceae.

In the Florideophyceae, apart from the 'monospores' (p. 136) detected recently in culture studies, all spores are produced in well defined and highly characteristic sporangia.   The major problem in defining what is meant by a spore in the many genera of the Bangiophyceae results from the fact that such a clear distinction between a vegetative cell and a spore or spore producing cell is usually not the case in this group. By gathering together the scattered literature, Drew (1956) was able to present a classification of spore types in the Bangiophyceae based on their manner of formation.   She was able to distinguish three types (Fig. 29):

Type 1 spores were monospores produced by differentiated sporangia;
Type 2 spores were monospores produced from undifferentiated cells;
Type 3 spores were spores produced in some numbers by the successive divisions of a mother-cell.

Considering first the two types of monospore which were distinguished, the crucial difference between these relates to the level of differentiation. If a cell underwent no differentiation whatsoever, so that every cell division prior to the release of a spore was identical to those characteristic of vegetative growth, the spore was said to be of Type 2.   Such a spore represents simply the release from a multicellular body either of a whole cell or the contents of a cell.   Type 1 spores were those where cell function changed immediately prior to the formation of a sporangium, in that a specialised portion of a cell was cut off, by a division different to those by which vegetative growth was achieved. A spore was then released from this specially cut off portion, leaving behind a larger sterile cell.   In general, it has been customary to regard monospores of either Types 1 or 2 as not being of sexual origin or related to a sexual process (Drew, 1956).

Such an interpretation would appear to be correct for the Type 2 monospores reported in *Asterocytis, Bangia, Erythrotrichia, Goniotrichum, Kyliniella, Neevea, Porphyra* and *Porphyrella*.   Because of the absence

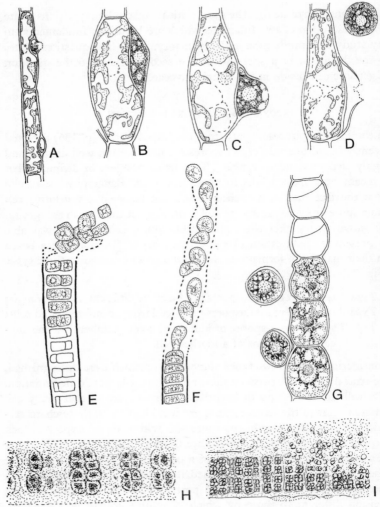

Fig. 29. *Spore and gamete production in Bangiophyceae.* A–D, Rhodochaete parvula; A, *young and mature Type 1 'spermatangia', the latter in the act of release* (×750); B, *young Type 1 'sporangium'* (×750); C, *mature Type 1 'sporangium'* (×750); D, *discharged Type 1 'sporangium' and 'spore'* (×750); E, F, Bangia fuscopurpurea, *shedding of Type 2 'spores'* (×300); G, Erythrotrichia carnea, *production and release of Type 1 'spores'* (×540); H, Porphyra *sp., one form of Type 3 'spores', the so-called 'carpospores'* (×150); I, Porphyra *sp., one form of Type 3 'spores', the so-called 'spermatia'* (×150)

(A–D, *after Magne;* E, F, *after Drew;* G, *after Taylor;* H, I, *after Thuret and Bornet*) of any specialised differentiation prior to the formation of such Type 2 monospores, they have been frequently overlooked because only by prolonged observation of living material can such structures be seen. Although recorded by various authors in *Porphyra*, Drew (1956) concluded that Type 2 spores are 'either rare in the genus *Porphyra* or consistently overlooked'. In fact, such Type 2 monospores are frequently formed and in quantity by the old basal parts of thalli which are perennating under a covering of sand and debris. They can also be detected in fully-grown thalli, if these are observed for a time in the living state, being released, particularly from the margins of the frond. By comparison, the formation of such monospores has been seen and reported more often in *Bangia* (Fig. 29, E, F) than in *Porphyra*, particularly from thalli in the uniseriate condition. There is some doubt as to whether the spores released from the multi-seriate thallus are actually of Type 2. In *Goniotrichum*, on the other hand, Type 2 monospores are released from thalli in both the uniseriate and multiseriate states. As mentioned previously (p. 48) the thallus of *Asterocytis*, although usually filamentous in appearance, is difficult to distinguish from some of the colonial coccoid unicells and the formation of Type 2 monospores, by the release either of a cell or its contents, is very difficult to distinguish from the breakdown of a colonial organisation. Type 2 monospores have also been described in *Neevea*, *Porphyrella* and *Kyliniella* although a very different process has been described in the American material of the last genus (Flint, 1953). Although spore production in *Erythrotrichia* is usually of Type 1 monospores there is one reported exception for *E. welwitschii* (Dangeard, 1949) where the release of the entire contents of a cell as a spore has been described.

There are a number of unanswered questions concerning Type 2 spores. One of the most serious is that little is known about the fine structure of these spores or the sporangia in which they are formed, and this deficiency is even greater for the other two types. Another problem concerns the mode of release and there might well be more than one mechanism. In uniseriate thalli, the spores are released at the apices by some form of general dissolution whereas in flat thalli, such as *Porphyra*, or in multiseriate thalli, the cells or cell contents are shed laterally with no trace remaining afterwards of the passage of the spore through the wall. In general, it has been customary to consider the release of Type 2 spores simply as the shedding of a cell or its contents from a multicellular thallus, although, as will be seen, the one fine structural investigation of Type 2 sporangia and spores indicates that this is not the case. A related problem concerns the nature of the outer layer of these Type 2 spores at the time of release. In some cases it would appear that a cell and at least the greater part of its original cell wall is released whereas in other cases the spore is described as 'naked' which implies that the limiting layer is relatively thin and probably not rigid. Amoeboid movement of Type 2 monospores has been reported in a number of species. By comparison, some investigators (Wille, 1900; Rosenvinge, 1909; Waern, 1952) have reported the release of extremely thickwalled Type 2 spores and it has been suggested that the term 'akinete' should be applied to these. There are thus several different sorts of limiting layer reported for spores of Type 2 and these cannot be evaluated without several fine structural investigations.

To date, there has been only a single study of the formation and release of Type 2 spores, in *Smithora naiadum* (McBride and Cole, 1969). This study disclosed that in this member of the Bangiophyceae, at least, the formation and release of Type 2 monospores involved more than a mere shedding of vegetative cells from a multicellular thallus. During the initial stages, a vegetative cell diminishes in size and the protoplast rounds off. At the same time, the dictyosomes gave rise to fibrillar vacuoles as well as to other vacuoles which are smaller than the fibrillar vacuoles and more transparent. The fibrillar vacuoles are similar to those which have been detected during both spermatogenesis and tetrasporogenesis in certain genera of the Florideophyceae (see p. 126). As in spermatogenesis, the fibrillar vacuoles observed in the monosporangia of *Smithora* appear to be involved in the production of mucilage. A further observation of interest in connection with the previous comments regarding the presence or absence of a wall in Type 2 monospores is that in *Smithora* the released spore is bounded only by a single plasmalemma and there is no trace of a cell wall or of any remains of the parent cell wall. This one investigation has provided an answer to many outstanding questions but it would be advisable to obtain confirmation of these observations in other entities before accepting them uncritically as being representative of Type 2 monosporangia throughout the Bangiophyceae.

Turning next to the Type 1 monospores, these are formed in differentiated sporangia, a process very different from the manner of formation of the Type 2 monospores.

The formation of a sporangium even in algae with such a relatively low level of differentiation as the Bangiophyceae is a very distinctive process. Following division of the nucleus, an oblique curved wall is laid down in the apical pole of a cell to divide it into two, unequal portions. Each portion contains a chloroplast and pyrenoid. The smaller portion is the spore producing sporangium, and, after release of its contained spore, the larger portion may expand to fill the space which it occupied and then divide again to form a second sporangium. Unlike the situation in the Florideophyceae (see p. 126), where the remnants of old sporangial walls can persist, no obvious traces of the preceding sporangium remain in these instances of the regeneration of a Type 1 sporangium in the Bangiophyceae. The most detailed inform- ation on these Type 1 sporangia is in the genus *Erythrotrichia* (Fig. 29, G) where they may be detected with relative ease, both on filaments of the upright thallus and also on the basal disc where this is present. *Erythrocladia* and *Erythropeltis* are said to resemble *Erythrotrichia* (Rosenvinge, 1909) although the information available on spore forma- tion in these genera is far from satisfactory. The production of mono- spores by a Type 1 procedure was used as the basis for the segregation of *Porphyra coccinea* as a distinct genus, *Porphyropsis* (Rosenvinge, 1909). In *Compsopogon*, spores are formed in two distinct ways, the products being of different sizes. The so-called 'macroaplanospores' are solitary and formed in a manner very similar to the formation of Type 1 mono- spores of *Erythrotrichia*. The 'microaplanospores' are smaller and formed in clusters either from the uncorticated uniseriate axes or from the cortication. Details of the development of the 'microaplanospores' are somewhat obscure, although it has been indicated that in *C. coeruleus* they may be produced in several very different ways (Nichols, 1964a). The interpretation of the 'macroaplanospores' and 'micro- aplanospores' has varied considerably. The 'macroaplanospores' have been regarded as lateral branches with arrested development (Skuja, 1938), whereas it has been suggested recently that there are no real differences between the two types of sporangium (Nichols, 1964a). In *Rhodochaete* (Bornet, 1892), the original description gave details of the formation of monosporangia which were formed in a manner somewhat different from those of *Erythrotrichia*. The sporangia were formed by a *lateral* curved wall rather than an acropetal inclined wall. In a more recent study (Magne, 1960) it has been shown that there were sporangia of two sizes which developed in this lateral manner, one about one third the size of the other. It was shown that the small cells functioned as male gametes, whereas the large spores were formed by cells with which the smaller 'spores' had fused previously. Thus, the Type 1 sporangia produced structures which must be regarded as male gametes and as post-fertilisation structures. In *Erythrotrichia* and several related genera, similar confusion occurs in the literature. Structures identical with the Type 1 spores, although somewhat smaller, were said to function as male gametes in the genera *Erythrotrichia*, *Erythropeltis* and *Erythrocladia*. It should be pointed out that the records are both few in number and extremely incomplete, being confined largely to the data of Berthold (1881, 1882b) with a few later observations. The situation with respect to the Erythropeltidaceae has been summarised as being in need of a 'thorough investigation particularly with regard to sexual reproduction' (Drew, 1956). Spores formed in differentiated sporangia are liberated by rupture of the sporangium wall and it has been suggested in various genera that the process of expulsion is violent, although in *Compsopogon* the macroaplanospores are said to glide from the sporangium. The liberated Type 1 spore, unlike the Type 2 and Type 3 spores, has never been said to undergo amoeboid movement or change of shape following liberation although 'conspicuous gliding movements' (Rosenvinge, 1909) have been reported for the Type 1 spores liberated by *Erythrotrichia carnea* and *E. reflexa*. Although the literature (Drew, 1956) implies that Type 1 spores are not of sexual origin or participate in the sexual process, the data suggests various examples which are contrary to this hypothesis. Either *Erythrotrichia* produces Type 1 sporangia of two different sizes or one is a male gamete and the other an asexual spore (see p. 177); the evidence on these points is far from clear. It has been suggested, on even more slender evidence, that the 'microaplanospores' and 'macroaplanospores' of *Compsopogon* both function as gametes (see p. 174) and fuse. The most significant evidence is in *Rhodochaete* (Magne, 1960) in which there are two Type 1 'spores', one smaller than the other. The small 'spore' functions as a male gamete (Fig. 29, A), which fuses with an undifferentiated cell. Following the fusion with the smaller 'spore', the fusion cell divides to form the larger 'spore' (Fig. 29, B-D). Thus, the hypothesis regarding the Type 1 spores as 'asexual' must be rejected; their function is far from clear in that in addition to these instances in which there is clear evidence of the function of a Type 1 spore as a male gamete, or as a post-fertilisation spore, there is equally good evidence in a few cases that they do in fact function as 'asexual' spores (see p. 210).

The final type of spore formation (Type 3, of Drew) occurs through the successive divisions of a mother cell.   Such divisions, by walls laid down in two or three planes at right angles, have been known for many years in species of *Porphyra* and they have also been reported on many occasions in *Bangia*.   There has been some confusion regarding the latter genus, probably as a consequence of the simultaneous formation of more than one spore type on the same thallus.   In both genera the volume of the mother cell increases only slightly during the process of division while the original cell wall remains distinct so that the mature spores are arranged in obvious 'packets'.   Both in *Porphyra* and *Bangia*, two different forms of Type 3 spores may be produced.

In one case, each mother cell usually divides two or three times to give either 4 or 8 products, although there may be a larger number formed in certain instances by further division.   It is usually accepted that 8 spores are formed from each mother cell in species of *Porphyra* (Fig. 29, H) although the degree of variation is probably greater than has been accepted hitherto while only 4 spores per mother cell arise consistently in some species.   There appears to be less consistency in the number of products formed from each mother cell in *Bangia* so that 4 or 8 spores may be produced by adjacent mother cells.   There is some disagreement as to the orientation of the initial division while the interpretation of the function of these bodies has been related to the plane of the first segmentation.   Some authors have considered the first division to be parallel to the surface of the thallus although Hus (1902), in a very detailed treatment of this, claimed that both the first and second divisions were at right angles to the surface, in this way giving the very characteristic cruciate aspect to this stage when seen in surface view.   The crucial feature is that the products remain pigmented and are released in this condition.   The function of both the mother cells and the products of their division has been the subject of much controversy.   The mother cell has been considered as a gamete (see p. 176), which is fertilised although the spores have also been considered as gametes (see p. 174).   Some authors differentiate between the products of division of the mother cell according to whether the first division is parallel to or at right angles to the surface of the thallus (Berthold, 1882b), the former being considered as the products of division of a fertilised gamete whereas the latter are said to be asexual spores.   In general, most authors have considered these bodies to represent the product of some form of sexual reproduction and referred to them as 'carpospores'.

In the second case, the mother cell divides to produce a much larger number of products (Fig. 29, 1). There is usually said to be 64 in most species of *Bangia* and *Porphyra*, although in the latter there may be as many as 128 or as few as 16 or even 8 in some of the distromatic species. However many products are formed, pigment is lost by the chloroplast during the process of division although the rate of loss varies from species to species. The chloroplast also decreases in size but it may be recognisable in the liberated product which is usually colourless. As these are formed from large numbers of mother cells, areas of the thallus may be pale yellow or greenish. Because of the greater number of divisions and the minimal increase in size of the mother cell during division the resulting colourless bodies are much smaller than the pigmented type discussed previously. There is the same disagreement found with the latter regarding the initial division which produces the colourless products, although the majority of authors consider that the first division is parallel to the surface of the thallus. The products are liberated in vast numbers by the dissolution of the cell walls and have been reported as exhibiting amoeboid movement. Early reports of motility are now considered to represent a fungal infection. Virtually all authors have considered these smaller, non-pigmented bodies to be male gametes and they have been referred to as 'spermatia' in consequence.

Although these supposed 'carpospores' and 'spermatia' of *Porphyra* and *Bangia* can be obtained in vast quantities with ease, the evidence on which such interpretations are based is virtually non-existent. About all that can be stated with certainty is that *Porphyra* and *Bangia* produce spores of two different sizes by the repeated division of mother cells. Obviously one should not continue to refer to these as carpospores or spermatia. The nomenclature to be used has been considered by Conway (1964) who has suggested that it would be most appropriate to refer to these simply as large spores, or α-spores, and small spores, β-spores. Despite the variation in number of divisions in both cases, and the resulting variation in size, the two classes are quite distinctive and can usually be recognised simply on the basis of size and colour.

Spore formation by a process similar to the successive divisions of a mother cell in *Bangia* and *Porphyra* has been reported also in the little known *Phragmonema sordida* first detected growing epiphytically in the orchid house at the Botanical Garden, Berlin. Cells of the simple thallus divided by longitudinal and transverse walls to give 8 or more bodies which were released by dissolution of the mucilaginous cell wall.   The equally obscure *Cyanoderma bradypodis*, which inhabits the hairs of the sloth, behaves in a similar manner, any cell dividing endogenously to give as many as 40 products, as does the unicellular *Rhodospora sordida* which forms several spores per cell through endogenous division.   The bodies are released in the two latter cases by dissolution of a part or the whole of the cell wall.   It is possible, although somewhat unlikely, that the various reports of tetrasporangia in members of the Bangiophyceae should also be treated as variants of Drew's Type 3 sporangia.   Structures in which the contents were divided into four parts have been described for *Compsopogon* (Flint, 1947; Das, 1963) and *Kyliniella* (Flint, 1953) although the documentation is extremely unsatisfactory so that in no case was it even established that spores were actually released, much less that they were capable of germination or played any part in the life history.

In general, the spores of *Porphyra* and *Bangia* referred by Drew to Type 3 have been regarded by most investigators as being derived from a zygote although with some suggestion that in certain cases, depending on the plane of the initial division, the spores may not result from a sexual process.   Also, there has been little agreement as to whether the first division of the nucleus was mitotic or meiotic.   The remaining examples of this type of spore formation reported in genera other than *Bangia* and *Porphyra* are of unknown significance.

In summary, it can be stated that unicellular bodies of different shapes, sizes or forms may be produced in members of the Bangiophyceae in one or more of three different ways.   The method of formation of the spores may be a characteristic of the entity under consideration in many cases.   Although the three methods of formation outlined by Drew (1956) provide a useful categorisation, bodies of several different types may be included in each.   The major difficulty is that there is little real understanding of the origin, behaviour and function of the spores referred to Type 1 and Type 3.   In both cases, examples occur which are clearly derived from a zygote while other cases are clearly not preceded by a sexual process.   In addition, there is no obvious clear-cut distinction between spores and gametes.   As will be seen later in the treatment of life histories in the Bangiophyceae, the patterns of germination of these various spores also vary and any arrangement of spore types in the Bangiophyceae must take into account the pattern of germination as well as the criteria employed by Drew, namely origin, behaviour and function.   The question of sex in the Bangiophyceae is treated in more detail in the subsequent section while the data currently available on reproduction and reproductive processes is assembled later (see p. 205) into such a picture of the life histories as is possible at the present time (p. 210).

(see p. 205)
(p. 210)

THE PROBLEM OF SEX IN THE BANGIOPHYCEAE

Evidence of sexual reproduction consists in the recognition of gametes and observation of their fusion. Subsidiary evidence may be obtained from cytological studies indicating counts of the number of nuclei or chromosomes. Sexual reproduction has been reported for the following genera of the Bangiophyceae: *Bangia, Compsopogon, Erythrocladia, Erythropeltis, Erythrotrichia, Kyliniella, Porphyra, Porphyrella, Rhodochaete* and *Smithora*. In almost every case, the evidence on which these reports are based is very incomplete and highly conflicting. An attempt will be made to assess the data which exist and to evaluate the evidence on which the occurrence of sexual reproduction is claimed for the Bangiophyceae. This assessment of the data will consider the various forms of gametes which have been reported to participate in fusion, the relationship of these with the supposed gametes which have been reported on purely descriptive morphological grounds, the cytological evidence of fusion and post-fusion behaviour, and the consistency with which certain morphological developments are said to follow or be a consequence of sexual fusion.

No general statement can be made on the gametes reported in various genera and species of the Bangiophyceae. The descriptions may be grouped into four types, which can be summarised as follows. In each case a cell released from the thallus will simply be referred to as a 'body' in order to avoid any implication of function.

1. The fusion of two released 'bodies' has been described for *Porphyra* (Derbés and Solier, 1856; Koschtsug, 1872) and *Smithora* (Knox, 1926, as *Porphyra naiadum*) and suggested for *Bangia* (Reinke, 1878) and *Compsopogon* (Thaxter, 1900; Krishnamurthy, 1962). It was suggested (Goebel, 1878) that the occurrence of spores with protuberances could be brought about by amoeboid movement rather than by a process of fertilisation and similar observations, interpreted as budding, have been reported for *Porphyridium* (Sommerfeld and Nichols, 1970a) and *Rhodella* (Paasche and Throndsen, 1970). In *Porphyropsis coccinea* (Murray *et al.*, 1972 it was shown by time-lapse photography to be division

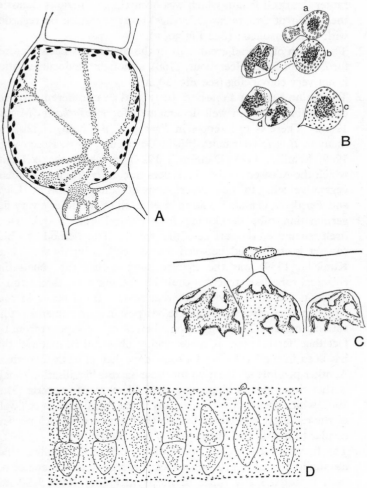

Fig. 30. *Alleged modes of gamete fusion in Bangiophyceae.* A, Ery-throtrichia carnea, *fusion of two adjacent cells in thallus* (× 1400); B, Smithora naiadum, *stages in the fusion of two released 'spores', showing these immediately prior to fusion (a), after fusion (b, c), and adjacent non-fusing cells (d)* (× 500); C, Porphyra *sp., fusion of a released 'spore' and an undifferentiated receptive cell* (× 1100); D, Porphyra perforata, *fusion of a released 'spore' and a cell with specialised receptive protuberance* (× 350)

(A, *after Heerebout;* B, *after Knox;* C, *after Berthold;* D, *after Smith*) rather than cell fusion which was occurring. Previous claims of the fusion of two released 'bodies' must therefore be regarded with some suspicion (See Fig 30, B).

2. The fusion of two adjacent cells in the thallus has been reported for *Erythrotrichia* (Heerebout, 1968), although the documentation is not very convincing (see Fig. 30, A).

3. The fusion between a released 'body' and an undifferentiated cell of a thallus has been well documented for *Rhodochaete* (Magne, 1960). There is one report in *Porphyra* (Berthold, 1882b) but many in *Bangia* (Berthold, 1882b; Darbishire, 1898; Rosenvinge, 1909; Schiller, 1925; Dangeard, 1927) in which the cell with which the released 'body' fuses does not differ from an ordinary vegetative cell. In these descriptions of gamete fusion in *Bangia* and *Porphyra*, contact is said to be effected by means of a very fine germination tube developing from the released 'body', which itself remains on the surface of the thallus. The method by which contact is established between the two cells is highly suspicious. Kunieda (1939) was the first to suggest that the penetrating rhizoidal tube-like structure could be nothing more than a fungal infection and it is now known that several fungi occur in *Porphyra*, in some cases sufficiently to be a pest in the commercial cultivation grounds. This suggestion receives strong support from the fact that 'fertilisation' is more readily observed in material that has been in the laboratory for some days than in fresh collections. Another possible explanation for these narrow 'fertilisation canals' is that these are nothing more than filamentous bacteria (Hus, 1902) which occur almost universally in the mucilaginous cell walls of the red algae and which frequently form narrow lines perpendicular to the surface of the thallus (see Fig. 30, c).

4. The fusion of a released 'body' with a cell of the thallus which has differentiated to produce a special receptive protuberance has been reported many times in *Porphyra* (Joffé, 1896; Ishikawa, 1921; Dangeard, 1927; Kunieda, 1939), but rarely in *Bangia* (Yabu, 1967). These observers reported the occurrence in *Porphyra* of protuberances of varying length which they interpret as being equivalent to the trichogynes of the Florideophyceae. The statements about these protuberances are extremely variable, with respect to the shape of the structures, their variability, and the age of those parts of the plant in which they are said to occur. The similarities between many of these supposed 'trichogynes'

and the consequences of *in situ* spore germination are such as to demand very close scrutiny of the former reports (Richardson, 1969) (see Fig. 30, D).

Reports of gamete fusion in *Erythrotrichia*, *Erythropeltis* and *Erythrocladia* are even fewer in number and, if anything, even more confused than for *Bangia* and *Porphyra*. In these genera, the reports are superficially similar to those in *Porphyra* in which a released 'body' is said to fuse with a superficial 'trichogyne-like' protuberance. The fusion of such a released 'body' with a specialised cell of the thallus has also been reported in *Kyliniella* (Flint, 1953); in fact, two different forms of gamete fusion of this type were said to occur in the same specimens. In one, two adjacent cells were involved, of which one lost turgidity and extended through the outer wall to give rise to a non-pigmented spore, while the adjacent cell enlarged considerably. The released 'body' was said to be liberated in proximity to a receptive area in the enlarged cell and fertilisation was effected. The second method was essentially similar except that it did not involve cells which developed adjacent to one another.

Evidence of a relationship between the released 'body' or 'bodies' reported as participating in fusion and the alleged gametes described on purely morphological grounds is minimal. In *Porphyra* and *Bangia*, colourless spores of Type 3 are produced in quantity in some specimens (see p. 171); these have been interpreted as male gametes and referred to as 'spermatia' by most workers. Despite their relatively widespread occurrence and production in considerable quantity under certain circumstances, there is no proof that these cells are the bodies which have been reported as participating in fusion. The spermatia are of similar size to some illustrations of alleged fusion but not of others. In the genera *Erythrocladia*, *Erythrotrichia* and *Erythropeltis* the described 'spermatia' are relatively rare and they are said not to be colourless bodies, as in *Porphyra* and *Bangia*, but to contain a recognisable pigmented chloroplast. Their interpretation is further complicated by the fact that there appear to be, in these genera, two sizes of Type 1 sporangia, the larger producing what is usually interpreted as an asexual 'spore', the smaller producing the supposed 'spermatium' (Berthold, 1882b; Baardseth, 1941). While this interpretation is interesting, the factual evidence to support it, to justify the existence of two size classes of Type 1 sporangia, producing products with two different functions, is simply not adequate at the present time. The interpretation does bear some similarity to the data for *Rhodochaete* (Magne, 1960) where it has been shown that, in addition to the large Type 1 monospore mentioned in the original description and all subsequent comments on this genus, there is a smaller Type 1 spore, which is in fact the male gamete which has been seen to fuse with an unspecialised vegetative cell. In conclusion, various 'spores' have been interpreted as male gametes, in almost every case, on extremely poor evidence.

The cytological evidence in support of fertilisation is scanty and contradictory even in such genera as *Bangia* and *Porphyra* which are well studied by comparison with other Bangiophyceae. At the nuclear level, observations of cells containing two nuclei are rare. Such binucleate cells have been reported or figured for *Bangia* (Dangeard, 1927) and *Porphyra* (Joffé, 1896; Dangeard, 1927; Magne, 1952; Tseng and Chang, 1955) but it should be appreciated that bi- or tri-nucleate cells have been shown to be produced in the latter genus under the influence of sewage discharges of industrial origin (see p. 43). It was argued by some workers (Dangeard, 1927; Tseng and Chang, 1954, 1955) on the basis of very scanty evidence that meiosis could be demonstrated in *Porphyra* prior to the division of the mother cells to form the so-called carpospores, whereas others (Magne, 1952; Yabu and Tokida, 1963; Kito, 1966, 1967; Kito *et al.*, 1967) claimed that such divisions were exclusively mitotic. A similar claim was made for division of the mother cells in *Bangia fuscopurpurea* (Yabu, 1967). In addition, some workers have indicated that the same chromosome numbers occurred at all stages in both *Porphyra* (Krishnamurthy, 1959) and *Bangia* (Richardson and Dixon, 1968; Sommerfeld and Nichols, 1970b). On the other hand, recent Japanese studies (Yabu, 1969, 1970, 1971) indicate that in the foliose thalli of various species of *Porphyra*, mother cells which are said to be those which give rise to the 'carpospores' have nuclei with a chromosome count twice that found in cells which are thought to be those giving rise to the 'spermatia'. The only other position in the life history of *Porphyra* where meiosis has been reported is in the fertile cell rows of the *Conchocelis* phase (Giraud and Magne, 1968). Thus, in *Porphyra* and *Bangia* there is little agreement as to cytological events from which the occurrence or absence of fertilisation might be deduced, and all that can be said is that doubling of the chromosome number may occur in certain circumstances although not indisputably under all conditions.

One of the few other cases where the cytological evidence substantiates the idea that fusion has occurred is in *Rhodochaete* (Magne, 1960) where the indications are that a doubling of the chromosome complement can be demonstrated in the course of zygote formation while the subsequent formation of the single post-fertilisation spore is mitotic.

To summarise, there are claims for the formation of gametes and their fusion, and there are indications of the doubling of nuclei and subsequently of chromosomes, but for each piece of positive evidence available at the present time there is an equivalent and equally convincing piece of negative evidence. Thus, although it seems difficult to deny the occurrence of sex in the Bangiophyceae on the basis of the various claims which have been made it is very nearly as difficult to state categorically that sexual reproduction plays a significant role in the biology of the Bangiophyceae.

Obviously, a considerable body of detailed observations and highly critical investigations is needed before it will be possible to give a final evaluation for the question of the occurrence of sex in the group as a whole. In order to establish beyond doubt the occurrence of sex in the Bangiophyceae it will be necessary to show that the cells said to be gametes actually participate in fusion, that the recipient gamete then contains two nuclei which subsequently fuse, that the chromosome number is then doubled, and that the same cell is involved throughout. On the basis of the present evidence, there would appear to be several different types of gamete and of gamete fusion in the group so that each example will need to be evaluated separately. One of the major problems at the present time is that there is little direct connection between the different pieces of available evidence. For instance, it is not known that the detached cell which is said to fuse represents one of the same cells which are identified on morphological grounds as spermatia, neither is there good evidence that those cells of the thallus in which the double number of chromosomes has been reported are identical with the potential mother cells of future carpospore clusters or that they represent the recipient cells with which the detached cells fused. One particularly critical question concerns the infrequency with which fertilisation has been detected and reported, because on the basis of currently accepted tenets each mother cell which gives rise to a cluster of carpospores represents a fertilised recipient gamete. In many species of *Porphyra* every cell in a considerable area of the thallus functions more or less simultaneously as the mother cell for a cluster of carpospores which implies that there must be millions of fertilisations

G

taking place within a relatively short period and that the process is 100% effective. Why then is the process virtually impossible to demonstrate? In the thallus of any species of *Porphyra*, large numbers of the potential carpospore mother cells can easily be scanned in surface view and also in optical section simply by folding over the thallus and focusing on the fold and no indication of fertilisation can be seen. One possible hypothesis worthy of consideration is that the process of gamete fusion in a plant such as *Porphyra* and the division of a 'carpospore' mother cell to form a cluster of 'carpospores' are separated in time by several or even many vegetative cell divisions. In those species where large areas of the thallus are converted into clusters of the so-called spermatia and clusters of the so-called carpospores the segregation might well be established while the frond is relatively small. The thallus must obviously be larger in those species where the areas giving rise to the two types of structure are smaller and interspersed although the same principle could apply equally well. It is also possible that there could be no direct relationship between the formation of a particular type of reproductive body and a sexual process and that the intervention of the latter is a consequence of some external factor or factors.

It is highly probable that problems of this nature and even the larger problem of the occurrence of sex in the Bangiophyceae will only be solved by taking a completely fresh approach and literally by setting aside the data which exist at the present time until sufficient new observations, obtained by stringently critical methods, are available for the previous data to be properly evaluated.

4

Life Histories

Some general considerations of life histories in the algae

One of the characteristic features of cryptogamic plants is the development from a reproductive body of a morphological phase of different aspect from the phase on which that reproductive body was formed. Nowhere is this phenomenon displayed with greater diversity than in the algae. Despite innumerable treatments of algal life histories in recent years (Drew, 1955, 1956; Chapman and Chapman, 1961) and considerable advances in knowledge, there is still much confusion caused by three principal circumstances. First there is no real agreement among phycologists as to the terminology or nomenclature to be applied either to the phenomenon as a whole, or to the constituent parts, or to the different sequences which have been demonstrated for particular entities. Secondly, the evidence accepted as 'proof' of a particular sequence is often of dubious validity. Thirdly, it is not generally appreciated that many of the disagreements may be due to the fact that behaviour, even in the same species, can be so flexible. Growing under different circumstances or in different geographical localities the sequence of phases or the patterns of behaviour can be very different.

Considering first the questions of nomenclature, the phenomenon under discussion will be termed 'life history' in the present treatment. There are at least two other terms, 'alternation of generations' and 'life cycle', both used widely, which need to be considered and the reasons for rejecting them in favour of 'life history' stated. The term 'alternation of generations' represents a translation, strictly speaking a mistranslation, of the German *Generationswechsel*. The latter was first used by the Danish zoologist Steenstrup (1842) for the succession of phases seen in the life histories of various invertebrates. This usage is, in fact, very different from its subsequent application by Hofmeister

(1851) in his botanical studies and it would be more correct to equate Steenstrup's usage with strophogenesis. Hofmeister demonstrated that in mosses, ferns and conifers there was a sequence of spore-producing and gamete-producing 'generations'. It was natural for Hofmeister, observing in those plants a phenomenon superficially similar to that discussed by Steenstrup, to have applied to it the same descriptive term. However, the word *Generationswechsel* does not mean *'alternation* of generations', but rather *'change* of generations'. The incorrect translation was used first in connection with the English version of Steenstrup's text and it was understandable that when Hofmeister's great work was translated the same English word should be applied. In addition to 'alternation of generations' being an incorrect translation, the term is much more representative of the situation characteristic of the higher cryptogams, where there is an almost obligatory alternation between spore-producing and gamete-producing phases. In the algae there are several life histories which are of a similar type to this strict 'Hofmeisterian cycle', but there are many more which are not. From the evolutionary standpoint it makes little sense to consider algal life histories from the very restrictive standpoint of the comparatively conservative land flora. The transmigration to the land has eliminated much of the potential for flexibility in terms of structure, reproduction and life history, which are expressed with greater breadth in the aquatic environment. There are many life histories in the algae and only one of these has survived in the course of terrestrial colonisation. One cannot look back from the one to the many and try to encompass all the latter into the restricted scope of the life history characteristic of some bryophyte or other higher cryptogam. A further objection is that there are many algae with life histories which involve more than two phases, as in most Florideophyceae, and one cannot have an alternation of *three* objects. With respect to the term 'life cycle', this implies that the sequence of phases consists of a single, obligate cycle. In some algae the life history is not obviously cyclic at all, while in many cases it would be better described as polycyclic. For these reasons, 'life cycle' is hardly suitable a term for general algal use, even if it might be applicable for the higher cryptogams, and it is difficult to see why it continues to be used so extensively with the algae. Without question, the term 'life history' is the best to use for these phenomena, in that it is clear, precise and devoid of unnecessary or unwanted connotations.

The next nomenclatural problem concerns the names to be applied to the constituent parts of a life history. Here again, the major difficulties come from attempts to apply to the algae a terminology developed for the higher cryptogams. In the latter, except for instances of apogamy and apospory, there is a regular obligatory cyclic alternation of two morphological phases, one of which produces gametes, the other spores. Not only are these two phases morphologically distinct, but there is complete coincidence between the somatic phase and the nuclear state. The gamete-producing phase is haploid, the spore-producing phase is diploid, and the terms 'gametophyte' and 'sporophyte' have been applied to these. As a consequence of the uniformity in this respect throughout the mosses, liverworts, ferns and fern-allies it was inevitable that attempts should be made to apply the same terms, without question, to the algae. The position with regard to life histories in the algae is very different in that there is certainly no single uniform type but considerable diversity. Some algal life histories are comparable with those of the higher cryptogams although many are not. In particular, the coincidence between somatic phase and nuclear state breaks down extensively. In *Fucus*, for example, gametes are produced on a diploid plant and there have been disputes as to whether this should be considered as a gametophyte or a sporophyte. In many Florideophyceae, the carpospore-forming phase is diploid but is retained on the haploid gamete-producing phase while there is a second distinct diploid spore-forming phase producing tetraspores. Finally, in *Ectocarpus siliculosus* and *Spermothamnion repens*, etc., the life history contains several similar morphological phases of different ploidy levels, each of which may produce reproductive structures of two different types. As with the terminology for the phenomenon of life history as a whole, it would seem best to avoid terms such as 'gametophyte' and 'sporophyte' as they are certainly not applicable in the manner for which they were devised in the higher cryptogams. The simplest procedure, that which will be adopted in the present treatment, is to name each morphological phase in terms of the reproductive structure produced, so that one may speak of 'gametangial phase' or 'tetrasporangial phase' and where more than one type of reproductive structure is produced, as in *Spermothamnion repens*, a morphological phase may be both gametangial *and* tetrasporangial.

The third nomenclatural problem concerns the discrimination between different algal life histories. As the known diversity of algal life histories has increased, so has this problem. At the beginning of the present century, most of the work on algal life histories had been undertaken by continental European botanists working with freshwater

Chlorophyta. The general impression at that time was that *all* algae
had a life history like *Spirogyra*, where the plant body is haploid and
the diploid state occurs only in the zygote. A diploid morphological
phase was thought to occur only in the higher cryptogams where it
represented a new 'antithetic' part of the life history (Bower, 1908), the
means by which the transmigration to the land had been effected.
Studies of *Polysiphonia* (Yamanouchi, 1906) and *Scinaia* (Svedelius,
1915) showed that the life histories of these organisms differed both
from those of higher cryptogams as well as the life histories of the algae
known at that time. Svedelius (1915) devised a new terminology to
accommodate *Polysiphonia* and *Scinaia*, terming the latter 'haplobiontic'
in that he thought there was a single morphological unit (or 'biont') in
the life history, while *Polysiphonia* was termed 'diplobiontic' because
the life history involved two kinds of individuals. This scheme did
not achieve wide acceptance because it was too imprecise, particularly
with regard to the nuclear states. A later treatment by Svedelius (1931)
developed a second terminology for algal life histories, based essentially
on a consideration only of nuclear states. This considered three
categories, haplonts, diplonts, and diplohaplonts, defined as follows:

> *haplonts:* algae where only the zygote is diploid, with meiosis occur-
> ring at its germination,
> *diplonts:* algae where only the gametes are haploid, with meiosis
> occurring at their formation,
> *diplohaplonts:* algae in which both diploid and haploid morphological
> units occur.

This scheme was widely accepted, with various minor modifications,
and applied to all algal divisions even though not particularly suitable
for the Florideophyceae because of the presence of the carposporophyte.
Another series of schemes was then proposed, initially for the Phaeophyta (Kylin, 1938; Smith, 1938), where distinctions were made
between those algae in which the morphological phases were similar or
dissimilar. The terms 'isomorphic' and 'heteromorphic' were applied
to these categories. Subsequently, Fritsch (1942) made use of this
distinction and applied it, with certain additional data, to all algae,
including the Rhodophyta. Traces of this terminology have survived
and various authors apply the term 'isomorphic' to those Rhodophyta
in which the gametangial and tetrasporangial phases are morphologically identical, while those in which these two phases are of different
morphology are termed 'heteromorphic'. This extension of the

original terminology overlooks completely the carposporangial phase. If, as shown subsequently (p. 186), the carposporophyte is to be regarded as a morphological phase equivalent to the gametangial and tetrasporangial phases, then *all* Florideophyceae must be heteromorphic. The difference between the two major types of life history is in terms of the *degree* of heteromorphy, in that there are two dissimilar morphologies in some examples and three in others.

Later developments involved recognition that the life history of any alga, particularly of the Rhodophyta, *must* consider both the number of cytological states as well as the number of morphological phases. Several schemes involving both considerations were proposed by Hygen (1945), Chadefaud (1952), Feldmann (1952b) and Drew (1955). The views of Chadefaud and Feldmann are currently accepted by most French workers while those of Drew have been used by most English-speaking investigators. The scheme by Hygen was largely ignored. The present treatment is not the place to discuss the details of each scheme or the terminology which was developed. The final step in descriptive terminology was made by Chapman and Chapman (1961) who argued, quite correctly, that it was not sufficient to consider only the number of different morphological units but the total number of units in any particular life history is also significant. The only difficulty with this suggestion is that it adds yet another term to the description. For instance, *Polysiphonia* was described by Drew as having a dimorphic diplohaplont life history. With the additional information required by the Chapman and Chapman scheme, this becomes a dimorphic trigenic dibiontic diplohaplont life history. In an effort to simplify terminology, it was suggested (Dixon, 1963c) that a 'type method' was simpler and much clearer than the descriptive terminologies. The use of a 'type' nomenclature was not new, in that it had been partially adopted by Kylin (1938) and Drew (1944), but not completely. The adoption of 'types' is simpler than the descriptive terminologies which are becoming increasingly elaborate and specialised but it is also more flexible and capable of infinite expansion to accommodate new data which are obtained. As has been pointed out (Knaggs, 1970) the final list of 'types' may be long, but at least these will be simpler to express than the list of descriptive terms which would be required for these categories. Knaggs also suggested that the type should be indicated in terms, not merely of the generic names, but also of the species concerned in that variation of life history in a single genus is highly probable. While this argument has merit when the life histories of a sufficient number of species have been fully elucidated, it does not yet seem necessary unless there is more than one life history known in a genus used as a 'type'.

The life history of an alga may be defined (Drew, 1955) as the recurring sequence of morphological and cytological phases found in the species under consideration. A morphological phase may be defined (Drew, 1955) as a state of an organism recognisable by a constant characteristic morphological appearance irrespective of chromosome number. Basically, a morphological phase may be considered as beginning with a single cell, the spore or zygote from which it arose, and ending with a single cell, the reproductive body which it produces. In the Florideophyceae, the interpretation of the carposporophyte has long been the subject for controversy, although virtually all major authors (Kylin, 1937a, 1956; Smith, 1938, 1955; Fritsch, 1945; Drew, 1954a) have considered it as a phase, stage or generation equivalent to the structures which produce the gametangia or tetrasporangia. The most recent rejection of this concept appears to have been Svedelius (1927). As the carposporophyte begins with a single-cell (the zygote) and terminates with the production of one or more carpospores it fulfills all the requisites for a morphological phase and it is best that it be so interpreted. A cytological phase may be defined (Drew, 1955) as that state of an organism characterised by mitotic divisions all showing the same chromosome number.

In order to establish a life history certain specific data must be available. In view of the often doubtful evidence used as the basis for accepting a life history and the need for a highly critical approach, some detailed discussion of the required information is appropriate. The first set of data concerns a knowledge of the occurrence in the field of each morphological entity and reproductive body representing the species under consideration. This may or may not be easy to obtain. As will be seen, there are some taxa for which the major morphological units are identical, although in others they are completely dissimilar. The second set of data concerns the germination of each reproductive body and its growth to the point where the resulting plant is reproductively mature itself. Such information helps to associate morphological units which are of unlike appearance as well as establishing beyond doubt that the species does follow a certain sequential pattern of morphological phases. The third set of data relates to the cytological phases. It demands a knowledge of the chromosome number of each of the morphological entities involved as well as requiring the determination of the positions of syngamy and meiosis. These three basic requirements are sufficient to establish beyond doubt the life history as it occurs in the laboratory; the only difficulty is that for most red algae one or more of these requirements is lacking and the number of entities for which all the required data are available is extremely small. As a consequence, the lack of data in many cases has resulted in various assumptions being made and constant repetition of these obscures the critical fact that one is dealing with only an assumption. The final requirement is that the conclusions drawn from the acquisition of these data in the laboratory are confirmed in the field. As will be seen later (p. 210), there are often dramatic differences between laboratory-based conclusions and what is actually happening in the field.

This discussion of life histories in the red algae will consider first the Florideophyceae and then the Bangiophyceae. It will conclude with a discussion of the relationships between laboratory data and field observations, that is, of the theoretical and biological life histories.

Life histories in the Florideophyceae

Even though there was no understanding of their structure or function, it was appreciated by some of the earlier phycologists such as Stackhouse (1795-1801) that there are different types of reproductive structures in what are now considered as members of the Florideophyceae and that these occurred on different plants. During the nineteenth century, the pioneer work of Bornet and Thuret (1867) demonstrated gametangial fusion in several genera of the Florideophyceae and for the first time elucidated the nature of the carposporophyte and its relationship to the cystocarp. Strasburger published his generalisation on the periodic reduction of chromosome numbers in 1894 and this changed completely the outlook on life histories. In addition, the significance of both syngamy and meiosis was now apparent. For the red algae, the first demonstration of meiosis followed some years later, when Yamanouchi (1906) showed that it occurred in the tetrasporangia of *Polysiphonia flexicaulis* (as *P. violacea*).

The life history established for *Polysiphonia* by Yamanouchi, initially on grounds of cytology alone, has been substantiated by culture studies so that there is now a large number of cases where it is known that the life history consists of a sequence of gametangial, carposporangial and tetrasporangial phases, the first being haploid and the last two diploid (Fig. 31, p. 204). In certain genera such as *Pleonosporium* and *Tiffaniella* tetrasporangia are replaced by polysporangia. Drew (1937) demonstrated the occurrence of syngamy and meiosis in *Tiffaniella synderae* (as *Spermothamnion snyderae*) and showed that the gametangial plant is haploid while the carposporangial and tetrasporangial plants are diploid. Unfortunately the morphological sequence has not been proved by cultural studies in this species although demonstrated in *Pleonosporium borreri* and *P. dasyoides* for which cytological evidence is lacking. Assuming that one can integrate these data, this suggests that the life history of those Florideophyceae where polysporangia replace tetrasporangia consists of a sequence of gametangial, carposporangial and polysporangial phases, the first being haploid and the two last diploid. A further variant is found in *Plumaria elegans*. This alga is characterised by the occurrence of gametangial plants on which carposporophytes develop and tetrasporangial plants, as in *Polysiphonia*, but with the addition that there are also plants on which only parasporangia occur. Drew (1939) demonstrated that the gametangial plant was haploid and that syngamy occurred in the carpogonium; that the carposporangial and tetrasporangial plants were diploid with meiosis in the tetrasporangium; and that the parasporangial plants were triploid with only mitotic divisions in the development of parasporangia. On the basis of this cytological evidence alone, because no cultural data were available at that time, Drew postulated that in this alga there were two, completely independent, parts of the life history. On the one hand the gametangial, carposporangial and tetrasporangial phases constituted one life history which was identical with that of *Polysiphonia*. On the other hand the paraspores formed by the parasporangial triploid plant were presumed simply to repeat that triploid parasporangial phase. Proof of the last assumption has been obtained by Rueness (1968) although not for the *Polysiphonia*-like sequence of gametangial, carposporangial and tetrasporangial phases. Despite the abundant data confirming the occurrence of the *Polysiphonia* life history in many Florideophyceae, various kinds of evidence are also available indicative of deviation from a strict *Polysiphonia* life history in various algae of that class. These include reports of the occurrence of both gametangia and tetrasporangia on the same thallus, of the germination of tetraspores to give tetrasporangial plants and of the cytological demonstration for non-meiotic division in tetrasporangia.

Irregularities in the arrangement of reproductive structures characteristic of a strict *Polysiphonia* type of life history have been reported for many years. Detailed catalogues of these reports have been provided by Kniep (1928) and Knaggs (1970). One of the most curious features is that this phenomenon is more prevalent in the Ceramiales than in any other order, and of the Ceramiales most instances occur in the one family, Ceramiaceae. It is to be regretted that to date there have been so few cytological or cultural investigations. Drew (1934, 1943) showed in *Spermothamnion repens* (as *S. turneri*) that gametangial plants with approximately 30, 60 and 90 chromosomes occurred, although the procarps formed on the last degenerated. Carposporophytes with approximately 60, 90 and 120 chromosomes were detected, while tetrasporangia with meiosis were demonstrated in the 60-chromosomed plants. In addition, plants with 45 chromosomes were detected suggesting that these might have arisen by meiosis in tetrasporangia of the 90-chromosomed plants. Although Drew's conclusions were questioned as to the interpretation of the various ploidy levels (Barber, 1947) the principal conclusions are not affected by this. In *Callithamnion corymbosum*, Hassinger-Huizinga (1952) indicated that both tetrasporangia and gametangia might occur on the same thallus and it was claimed that the expression of gametangia and tetrasporangia is governed by the addition or deletion of a chromosome. These data must be regarded with caution since neither figures nor photographs of the chromosome complements are given and the results necessitate absolute accuracy in counting chromosomes with a complement of about 60 in a nucleus 5 μm in diameter. This requires technical ability of an order hitherto impossible in the Rhodophyta.

Germination of tetraspores to give tetrasporangial thalli has been postulated for many years as an explanation of the life histories of those species for which tetrasporangia were the only known reproductive structure. The work on *Callithamnion corymbosum* by Hassinger-Huizinga (1952) included studies of the germination of tetraspores; she showed that some gave rise to spermatangial and carpogonial plants while other tetraspores germinated to give tetrasporangial plants. Sundene (1962) showed that in *Antithamnion boreale*, for which no gametangial plants are known in the field, the tetraspores germinated to give only tetrasporangial plants for three successive generations. Cultural studies reported by West and Norris (1966) are sufficient to indicate various deviations even if a statement of the total expression of the life history cannot be made for any particular example. Spermatangial plants of *Antithamnion occidentale* eventually formed tetrasporangia, the spores from which germinated to form male plants on which tetrasporangia were absent. Similarly, in *A. pygmaeum*, carpogonial plants on which carposporophytes were present formed tetrasporangia, the spores from which germinated to form female plants. Tetraspores shed by an unidentified species of *Callithamnion* developed into male and tetrasporangial plants in the ratio of approximately 1 : 3 and the tetraspores of this and subsequent generations repeated this phenomenon. Tetrasporangial plants of two species of *Fauchea* collected in the field liberated tetraspores which germinated to give plants which were either entirely tetrasporangial (*F. laciniata*) or produced a mixture of tetrasporangia and bisporangia (*F. pygmaea*).

This last observation on a mixture of tetrasporangia and bisporangia is a phenomenon relevant to a discussion of the cytology of tetrasporangia in relation to non-meiotic nuclear division. The occurrence of sporangia containing only two spores has been widely reported in the Florideophyceae (Bauch, 1937), particularly in the Corallinaceae and Ceramiales, although they have also been reported in a few representatives of the Nemaliales and Gigartinales. Bauch was able to show that the individual spores could be either uninucleate or binucleate but his speculations as to the relationship between those two types has little factual basis. Subsequently, Suneson (1950) in a study of *Lithophyllum litorale* and *L. corallinae* showed that in the former, for which only plants bearing bisporangia are reported, meiosis did not occur in the bisporangia. The original nucleus of the sporangial primordium divided once mitotically, the derivative nuclei passing into the spores which are thus uninucleate at maturity. The situation in *L. corallinae* is very much more complex. Only gametangial and bisporangial plants had been reported from Sweden prior to Suneson's investigation although tetrasporangial plants were known from the Mediterranean. Suneson was able to demonstrate the occurrence in Sweden, of three different types of structure, namely, quadrinucleate tetrasporangia, quadrinucleate bisporangia, and binucleate bisporangia—with meiosis occurring in the first two types of sporangium but not in the third. This is the only instance of facultative apomeiosis in the Rhodophyta for which the data are sufficiently sound. The occurrence of apomeiosis in *Lomentaria orcadensis* (called *L. rosea* in the older literature) was claimed by Svedelius (1937). The data presented are extremely curious in that it was claimed that in the tetrasporangium the chromosomes migrated into the nucleolus where mitotic division took place. Magne (1964a) has reinvestigated this species and shown that the nuclear divisions in the tetrasporangia are perfectly normal. He has suggested that Svedelius may well have misinterpreted granules within the nucleolus, the granules being extremely prominent on occasions. The life history of

L. orcadensis is still a matter of conjecture. It is usually considered that gametangial-carposporangial plants are not present in Europe even though reported from Japan (Segawa, 1936). The conspecificity of the European and Japanese material has been questioned. There are, nevertheless, two reports of the occurrence of gametangial plants in Europe, one (Crouan and Crouan, 1867) which shows a plant very similar in appearance to the Japanese plant, although the second report (Lodge, 1948) is based on a misidentification of the 'rough water' form of *L. clavellosa*. Magne's study of the cytology of *L. orcadensis* has also produced some very curious information in that chromosome counts during diakinesis of $n = 10$ and $n = 20$ were obtained in material from different localities. The significance of the data in relation to the life history is still unknown. In the case of *Rhodymenia palmata*, knowledge of reproductive structures and life history is far from complete despite its widespread occurrence, great abundance and economic utilisation. Tetrasporangia are formed on most plants and although spermatangia have been reported by numerous investigators carpogonia and carposporophytes are unknown, the record of carpogonia (Grubb, 1923a) being based on a misidentification of hairs. Cytologically, Magne (1959) obtained counts of fourteen chromosomes in all cells of the thallus and observed fourteen bivalents in the sporangia, the nuclei of which appear to undergo normal meiosis as suggested by Westbrook (1928) although the latter indicated a different chromosome count (> 20). Austin (1956) gave a count similar to that of Westbrook, while the count given in a study of the Pacific material referred to *R. palmata* f. *mollis* agrees with Magne. The only conclusion is that there are suggestions of apomeiosis in normal tetrasporangia, but no absolute cytological proof of its occurrence. That the presence only of tetrasporangia must be taken as proof of apomeiosis must be treated with extreme caution. There is an ever-increasing number of instances where one 'species' for which gametangial and carposporangial reproduction are alone reported has been shown to be part of the same life history as another species, often placed in a different genus, family, or even order, in which only tetrasporangia are known (see p. 194). It would appear that the kind of life history typified by *Polysiphonia* may not be as well established as it appears. Unfortunately, none of the many instances of possible deviation cited above has been complete. There is still a need for detailed studies of life history which take into account not only behaviour in culture but also behaviour in the field, together with cytological investigations of both types of material.

Following the work of Yamanouchi on *Polysiphonia*, obvious targets for investigation were those genera, particularly of the Nemaliales, for which tetrasporangial plants had not been reported, except doubtfully on rare occasions (Dixon, 1963d). Studies of *Scinaia* (Svedelius, 1915) and *Nemalion* (Kylin, 1916b) reinforced by later investigations of *Asparagopsis* and *Bonnemaisonia* (Svedelius, 1933), gave rise to the concept that in such plants meiosis followed immediately after syngamy, so that both gametangial and carposporangial phases were haploid. This claim was based on cytological evidence of an extremely doubtful nature and without any knowledge of the sequence of morphological phases based on cultural studies. Thus, the general idea developed that members of the Nemaliales were set aside from other Florideophyceae through this absence of tetrasporangial phases, a concept which became firmly entrenched in the literature despite the shaky evidence on which it was based. The first evidence to cast doubt on this generally-held concept came through cultural investigations. It was shown (Feldmann and Mazoyer, 1937) that the carpospores of *Bonnemaisonia asparagoides* germinated to give rise to plants which were identical with the alga previously described as *Hymenoclonium serpens* and this study was followed by later investigations which showed that carpospores of *Asparagopsis armata* germinated to give *Falkenbergia rufolanosa* (Feldmann and Feldmann, 1939a) while those of *Bonnemaisonia clavata* gave rise to a plant very like *Hymenoclonium serpens* (Feldmann and Feldmann, 1939b). The interesting features of these observations were that *Hymenoclonium* and *Falkenbergia* had been assigned previously to the Ceramiales and that tetrasporangia had been reported in field collections of both algae. The tetrasporangia were sometimes a little peculiar so that there was disagreement as to whether these structures should be interpreted as tetrasporangia or as 'buds' of the *Bonnemaisonia* gametangial phase arising on a prostrate 'protonemal' structure. Although opinions on this controversy finally settled in favour of the tetrasporangial interpretation, the most recent information indicates that there is good evidence for both situations to occur in a single species or even the same material of one species. The evidence regarding the members of the Bonnemaisoniaceae slowly appeared over 20 years. Plants of *Bonnemaisonia hamifera* were obtained by the germination of tetraspores of *Trailliella* (Harder, 1948) although it was initially reported that the tetraspores of *Falkenbergia* on germination, gave rise not to *Asparagopsis armata* but to further *Falkenbergia*. Chihara (1960) later germinated the tetraspores of *Falkenbergia hillebrandii* and showed that the germlings resembled *Falkenbergia* initially, although with further growth, differentiation occurred and the germlings came to resemble *Asparagopsis*. This longer growth period has been repeated for the tetraspores of *Falkenbergia rufolanosa* (Feldmann, 1965) and Chihara's results substantiated. By 1960, the sequence of morphological phases in which the gametangial and tetrasporangial phases were dissimilar, had been established completely for at least three taxa (*Bonnemaisonia asparagoides*/*Hymenoclonium*; *Bonnemaisonia hamifera*/*Trailliella*; *Asparagopsis taxiformis*/*Falkenbergia hillebrandii*) but the interpretations were still based on the previous cytological data, with the difficulties of trying to reconcile these with the cultural observations.

Following these cultural studies, a re-examination of the cytological aspects of the life histories of these organisms, initiated by Magne, showed that the previously accepted information needed drastic revision. Magne showed in every case which he examined (*Bonnemaisonia*, *Nemalion*, *Lemanea*) that syngamy was not followed immediately by meiosis and that the carposporophyte was diploid.

The new data indicate that in various members of the Nemaliales the life history consists of the same sequence of gametangial, carposporangial and tetrasporangial phases as in *Polysiphonia* but with the difference that the gametangial and tetrasporangial phases are morphologically dissimilar. In addition, evidence is beginning to appear which indicates that in various genera of the Cryptonemiales and Gigartinales for which tetrasporangial phases had not been detected the life history is also of this pattern. There are still many outstanding questions in most of the cases which have been investigated, with respect to one or more missing pieces of information. The identity of the tetrasporangial phase, determination of the germination product of one or more of the spores in a life history, and particularly details of cytological aspects are all items for which there are more questions than answers.

Briefly though, the position may be summarised as follows. The product of germination of a carpospore may be of many different forms. In addition to the two freshwater genera, *Lemanea* and *Batrachospermum*, for which an association with a filamentous '*Chantransia*-stage' has been known for more than a century, there are many genera and species (*Cumagloia*, *Nemalion helminthoides*, *Pseudogloiophloea*, *Liagora farinosa*) for which the product of germination is very similar to a member of the *Acrochaetium*/*Rhodochorton* assemblage (Dixon, 1970). Other species give rise to growths which are more organised, forming prostrate branched systems of varying degrees of aggregation. The more loosely associated germlings were described as the '*Naccaria* type' while the more compact were termed 'discoid' (Chemin, 1937). This categorisation is convenient although not of fundamental significance because there can be enormous variation in the degree of compactness even in the same species. Carpospore germlings of the *Naccaria* type were first detected in *Bonnemaisonia asparagoides* and subsequently in *B. clavata* and, more recently, in many other genera such as *Acrosymphyton*, *Atractophora*, *Delisia*, *Dudresnaya*, *Naccaria*, *Ptilonia* and *Schimmelmannia*. Because *Bonnemaisonia asparagoides* was the first species for which it was shown that the product of carpospore germination resembled *Hymenoclonium serpens* it has been customary to state that the latter 'species' represented the tetrasporangial phase in the life history of the former. In view of the number of cases in which such a product of carpospore germination has been detected it would seem better not to equate the two algae but rather to say, simply, that the morphology of the tetrasporangial phase is of the '*Hymenoclonium*-type'. Similar situations occur many times in the red algae.

The hitherto undetected phases in the life histories of various members of the Cryptonemiales and Gigartinales are discoid and obviously heterotrichous in organisation. For some, it has been possible to relate these discoid phases with 'species' recognised in the field, although not in all cases. For genera of the Gigartinales, the phase produced by germination of the carpospores of *Halarachnion ligulatum* has been equated with *Cruoria rosea* (Boillot, 1965), while the carpospore germlings of *Turnerella pennyi* were referred (South *et al.*, 1972) to *Cruoria arctica*, a species very close to *C. rosea*. The discoid carpospore germlings of the Mediterranean *Neurocaulon grandifolium* bore tetrasporangia, although no identification of this tetrasporangial phase was made (Codomier, 1969). Conversely, West (1972) has shown that tetraspores from the crustose Pacific *Petrocelis franciscana* germinated to form basal discs from which arose erect fronds similar to those of *Gigartina agardhii* or *G. papillata*. For genera of the Cryptonemiales, the discoid carpospore germlings of *Gloiosiphonia capillaris* have been said (Edelstein, 1970) to bear a close resemblance to *Cruoriopsis hauckii*, as figured by Newton (1931), although the latter illustration is probably based on a misidentification. For two species of the Pacific coast, *Thuretellopsis peggiana* (Dixon and Richardson, 1969a; Richardson and Dixon, 1970) and *Pikea californica* (Scott and Dixon, 1971), the associated phases in the life history do not appear to have been recognised in the field, although for the former it did resemble *Erythrodermis* in some respects. Of the various entities mentioned in this discussion there is a very close correlation in ordinal assignment between the carposporangial plants and their associated phases. The suggestion made some years ago (Dixon, 1963d), that members of the *Acrochaetium/Rhodochorton* assemblage or of the non-calcified encrusting forms could represent undetected phases in the life histories of other Florideophyceae is now supported by a considerable body of evidence. In addition to these two morphological types, there are those algae where the phase produced by the germination of the carpospore is even more highly organised and conspicuous. As mentioned previously, the carpospores of various species of *Asparagopsis* germinate to give *Falkenbergia* while the carpospores of *Bonnemaisonia hamifera* produce a plant identical with *Trailliella intricata*. More recently, plants referable taxonomically to the latter have been obtained also by the germination of the carpospores of *Bonnemaisonia nootkana* (Chihara, 1965).

Despite the accumulating evidence of this range of morphological phases, hitherto undetected, occurring in the life histories of a wide range of genera, from many orders, there is still uncertainty as to the nature of these phases in a number of cases. The early controversy as to the relationship has been mentioned previously. Both *Falkenbergia rufolanosa* and *Hymenoclonium serpens* had been reported as bearing tetrasporangia, although these were occasionally of peculiar form, particularly in the latter species. It was claimed by some that these structures were sporangia, from which spores were released (Feldmann and Feldmann, 1946) and by others that they were not sporangia at all, but rather primordia which developed into gametangial plants. On the basis of the latter interpretation, the morphological *Hymenoclonium* phase would have to be regarded as a juvenile condition or protonema rather than as a tetrasporangial phase. The release of spores from the equivalent structures in *Trailliella* supported the orthodox interpretation that the structures in *Hymenoclonium* were sporangia rather than primordia. The situation was then confused by the claim (Feldmann and Feldmann, 1952) that a tetraspore of *Falkenbergia* on germination repeated the *Falkenbergia* phase. Such an explanation brought back into focus the protonema concept and, despite warnings (Drew, 1955) that it might be difficult to distinguish between young developmental stages of *Asparagopsis* and *Falkenbergia*, it was some years before it was proved conclusively that the apparent repetition of *Falkenbergia rufolanosa* by the germination of its tetraspores was due to premature termination of the experimental culture. At the present time there is a mass of additional information gathered during the past 25 years although in many respects the situation is as confused as it was then. In several instances it has been shown that the product of germination of a carpospore gave rise to tetrasporangia, the release from these of tetraspores has been demonstrated, the tetraspores germinated and their development followed to such an extent that the product may be identified and, in a few cases, followed to the point where it is reproductively mature. On the other hand, there are several equally well-documented instances where the plant formed by the germination of carpospores gave rise directly to what is identifiable on morphological, if not reproductive grounds as the gametangial plant. Thus, there are two pathways of development and both have been demonstrated for certain species. The plant formed by the development of carpospores produced tetrasporangia from which tetraspores were released to give, on germination, gametangial plants (or plants identified on morphological grounds as the gametangial phase) in various species of the Nemaliales (*Bonnemaisonia hamifera*, *Asparagopsis armata*, *A. taxiformis*, *Liagora farinosa*, *Pseudogloiophloea confusa*), Gigartinales (*Halarachnion ligulatum*) and Cryptonemiales (*Acrosymphyton purpuriferum*, *Thuretellopsis peggiana*, *Pikea californica*). There are, in addition, various reports where the occurrence of tetrasporangia was reported on the plant formed by the development of a carpospore with no additional information regarding the germination of the tetraspores, or, if that is available, the nature of the product into which the germinating tetraspore develops. Such reports are available for *Bonnemaisonia asparagoides*, *Naccaria wiggii* and certain species of *Nemalion*. The other pathway of development, whereby a plant regarded on morphological grounds as the gametangial phase arises directly from the plant formed by the germination of carpospores has been reported in several species of the Nemaliales (*Atractophora hypnoides*, *Bonnemaisonia asparagoides*, *Delisia fimbriata*, *Naccaria wiggii*, etc.) and the Cryptonemiales (*Pikea californica*, *Schimmelmannia plumosa*). It is interesting to note that the French workers, having disagreed strongly with the initial proposal by Kylin that the gametangial phase in *Bonnemaisonia asparagoides* arose by direct development from the product of germination of a carpospore, subsequently reported such a situation for this species.

Interpretation would be facilitated if detailed cytological data were available for the majority of examples of each of the two methods of development, particularly for those species where both methods have been reported. The crucial question concerns the occurrence and position of meiosis and still is largely unanswered. The studies of von Stosch (1965) and Ramus (1969b) on *Liagora farinosa* and *Pseudogloiophloea confusa*, respectively, indicate that meiosis occurs in the tetrasporangia in these two organisms. The evidence obtained by Ramus, both cytological and cultural, is particularly well documented and this investigation represents one of the most complete studies of the life history of any red alga. For those organisms where the plant identified as the gametangial phase arises directly as a bud on the product of germination of a carpospore, the only cytological data available is of *Lemanea mamillosa*. Magne (1967a, b) has claimed that meiosis occurs in the apical cell of the upright filament which produces the gametangial phase. Three of the resulting four nuclei were discarded in small lateral protuberances with one continuing as the origin of the gametangial phase. If correct, this situation represents the first instance in the Rhodophyta of non-sporangial, somatic meiosis. Photometric measurement of the DNA content has been used by Hurdelbrink and Schwantes (1972) to suggest that meiosis in an unidentified species of *Batrachospermum* occurs in a position equivalent to that suggested by Magne for *Lemanea*. The three cases where some cytological data are available may well indicate the relationship between the two pathways of development but it is not possible to state that complete answers are available to the cytological problems involved. Reports of the occurrence of both methods of development in the same species indicate that an organism may develop in one way or another depending on external conditions. It could well be that the famous illustration of *Platoma bairdii* (Kuckuck, 1912) in which a heterotrichous base is figured giving rise to an erect frond but also bearing tetrasporangia (Fig. 11, c) represents the occurrence of both pathways of development occurring side by side. It is possible that the difference between the two pathways of development is not as great as it appears. The basic steps involved in the formation of a tetraspore involve the formation of a new wall within the tetrasporangial mother cell, the division of nucleus and cell, and finally liberation. Failure to liberate can result in germination *in situ*, although it should be noted that in red algae this term has been used in two ways. It has been used to describe the germination of spores which have been released from the sporangium in which they were produced but retained within some sort of fruiting structure, as well as to describe germination within the sporangium itself. Most of the earlier reports are of the former category which is not relevant to the present discussion. However, that type of *in situ* development where the spores develop within the parent sporangium could be seen as the elimination of the last step in spore formation. The situation in *Lemanea*, as outlined by Magne, represents the additional elimination of new wall formation. Obviously, an investigation of fine structural aspects in addition to orthodox nuclear cytology is needed for this question to be resolved.

As might be anticipated there are numerous instances of discrepancies in the data. Chihara (1962) obtained plants regarded as the gametangial phase developing directly from the product of germination of carpospores in *Delisia fimbriata* while Levring (1953) has described plants from field collections, similar in aspect to the gametangial phase but producing tetrasporangia. The evidence for the occurrence of tetrasporangial plants is much more convincing than in *Ptilonia* where Lucas and Perrin (1947) state, rather ambiguously, that 'the tetrasporangia seemed to be in tetrads although they were not clearly divided'. As Chihara (1962) commented, further re-examination is necessary before this report can be finally accepted. In the case of *Halarachnion*, tetrasporangia develop on the encrusting product formed by carpospore germination in *H. ligulatum* whereas tetrasporangial plants, similar in aspect to the gametangial phase are known for *H. latissimum* (Okamura, 1934). Within the Bonnemaisoniaceae there appear to be many instances where field observations diverge from data obtained in the laboratory.

In all the cases discussed previously where the gametangial and tetrasporangial plants are of different morphology it is the tetrasporangial phase which is always the smaller and the less significant in appearance. There are now, however, two instances where a converse situation has been demonstrated. In *Acrochaetium pectinatum* (West, 1968) the tetraspores germinate to give gametangial plants which were small compared with the tetrasporangial phase. Similar results have been obtained with *Rhodochorton floridulum* (Knaggs and Conway, 1964), the difference in size between tetrasporangial and gametangial plants being even more marked. It was claimed, without any direct evidence being presented, that a cytological study indicated the occurrence of meiosis in the tetrasporangia. Subsequently Knaggs (1965) figured tetrasporangial plants of *R. floridulum*, bearing structures which he suggested were spermatangia. As the small gametangial plants formed by the germination of tetraspores have been collected in the field growing epiphytically on the parent tetrasporangial plants of *R. floridulum* (Dixon, unpublished) it could well be that this is the explanation for the statement.

Quadripartite carposporangia were reported in *Liagora tetrasporifera* (Børgesen, 1927) when it was thought that life histories in the Florideophyceae could be considered on the basis of a *Polysiphonia* type and a further type, exemplified by *Nemalion* at that time, where meiosis was thought to follow syngamy in the carpogonium. Despite the total lack of other data, the speculation generated by the discovery of 'carpotetrasporangia' has run rampant. Such divided carposporangia are now known to occur in many Florideophyceae. They have been reported in several species of *Liagora* and *Helminthocladia* and the type species of *Yamadaella* (Abbott, 1970), as well as in certain members of the Phyllophoraceae. West (1969) has recently given detailed descriptions of carposporophyte development in *Rhodochorton purpureum* and *R. tenue* where the carposporophyte gives rise to quadripartite sporangia. As West pointed out, there are several possible interpretations of these structures.

Despite the importance of a knowledge of the life histories of *Liagora tetrasporifera* and other species where the carposporangia are quadripartite, it is only recently that any data of significance have been obtained. Spores of *L. tetrasporifera* formed in the quadripartite carposporangia germinate to form a filamentous growth resembling a species of the *Acrochaetium/Rhodochorton* assemblage (Couté, 1971). This is similar to the product of carpospore germination in other species of *Liagora* such as *L. farinosa* (von Stosch, 1965) and *L. distenta* (Couté, 1971), where the carposporangia are not quadripartite. A thallus of what was identified on morphological grounds as the gametangial phase of *L. tetrasporifera* arose from the *Acrochaetium/Rhodochorton*-like filamentous growth. Tetrasporangia, such as have been reported on the product of carpospore germination in *L. farinosa* (von Stosch, 1965) and *L. distenta* (Couté, 1971) were not observed on the equivalent stage in *L. tetrasporifera*. It is unfortunate that no cytological data were obtained in this investigation. A cytological study of a newly discovered species of *Helminthocladia*, *H. senegalensis* (Bodard, 1972) showed that meiosis occurred in the carpotetrasporangia. Providing that one is prepared to amalgamate the results of several investigations of diverse genera it is possible to accept this life history as a 'type', equivalent to that of *Polysiphonia*. However, this should be accepted with caution, until a full investigation of a single species has been undertaken, because of the many examples known of 'direct development' and with the account of the somatic meiosis in *Lemanea* (Magne, 1967a, b) available.

The members of the Phyllophoraceae exhibit various patterns of development of reproductive structures. It is assumed that the life histories are likely to be somewhat specialised although the amount of real data is minute. Some of the species concerned possess reproductive structures distributed in the manner characteristic of *Polysiphonia* while in others the production of quadripartite sporangia in association with a carposporophyte is reminiscent of the situation in *Liagora tetrasporifera*. The interpretation of reproduction and life histories within the Phyllophoraceae has undergone several significant changes in the past century. During the early days, the pustule-like and sometimes colourless reproductive structures were regarded as independent parasitic genera of the Florideophyceae. That these represented part of the reproductive apparatus of the alleged host species was demonstrated (Rosenvinge, 1929) but even with this clarification the situation remained extremely confused. The interpretation of reproduction and life history in the Phyllophoraceae may be summarised as follows. For certain species, such as *Phyllophora crispa* (previously termed *P. rubens* or *P. epiphylla*), *P. membranifolia*, *P. nervosa*, *P. heredia* and *Stenogramme interrupta*, the distribution of reproductive structures is of the *Polysiphonia* type with tetrasporangia occurring on one plant and gametangia, which ultimately give rise to carposporophytes, on an independent plant similar to the other in appearance. Proof that the life history is actually of the *Polysiphonia* type either by cytological investigation or by spore germination, is, however, totally lacking. The genus *Gymnogongrus* poses many problems. It has been indicated (Phillips, 1925a; Doubt, 1935; Schotter, 1968) that the 30 or so species described may be divided into two groups, those which produce gametangia and an internal carposporophyte, and those which give rise to gametangia but then external proliferations or nemathecia containing tetrasporangia. Furthermore, a species placed in one group may have its counterpart in the other, so similar in appearance as to be indistinguishable on the basis of vegetative morphology. The two species of a pair may have the same or a different geographical distribution. In *G. linearis*, Doubt (1935) demonstrated the production of gametangia and the occurrence of fertilisation followed by the development of carposporangia, but claimed that the same chromosome number occurred in the gametangial plant and the germinating carpospores. The second species of the pair associated with *G. linearis* is *G. platyphyllus* and in this Doubt was able to demonstrate the presence of carpogonia, but not of fertilisation, and also the formation of tetrasporangial nemathecia from the auxiliary cell. The same chromosome number occurred in the cells of the vegetative thallus, the carpogonial branch and the germinating tetraspores. A somewhat similar account of the development of the tetrasporangial nemathecium in *G. griffithsiae* was given by Gregory (1934). *Gymnogongrus norvegicus* was thought to differ from the other species of the genus in that both types of reproductive structure appeared in the one species although Schotter (1968) has recently claimed that this species can be divided into two components. One of these has an internal carposporophyte and the other a tetrasporangial nemathecium. He has retained the epithet *norvegicus* for the latter naming the component which produces the internal carposporophyte *G. devoniensis*. There is still no acceptable data as to the identity of the product of spore germination and further cytological data are essential before any comments on the life history can be made. A tetrasporangial nemathecium similar to that of *Gymnogongrus griffithsiae* is formed in *Phyllophora truncata* (= *P. brodiaei*) although the stages of development are far from clear. The carpogonia are said in many cases to lack a trichogyne (Rosenvinge, 1929) while Kylin (1930a) claims that a high proportion appear to abort. The carpogonium is said to fuse with the auxiliary cell from which the resulting tetrasporangial nemathecium develops. The cytological evidence produced (Claussen, 1929), that meiosis occurred in the tetrasporangium, is most unsatisfactory although a recent study (Newroth, 1972) has indicated similar results although with a different chromosome number. Rosenvinge (1929) germinated the tetraspores and obtained thalli which were similar morphologically to the original plant after a period of about three years. Precise determination of the nature of these germlings was impossible since they were sterile. The new cultural and cytological evidence (Newroth, 1972) suggests that the generally accepted concept of the life history of *Phyllophora truncata* (= *P. brodiaei*) may well be correct. This may be regarded as the sequence of a haploid gametangial phase and a diploid tetrasporangial phase, the latter occurring parasitically or partially parasitically on the gametangial phase. *Schizymenia epiphytica* appears to be very similar in many respects to *Phyllophora truncata* (Smith and Hollenberg, 1943; Abbott, 1967) although nothing is known of it, cytologically or culturally and it is currently referred to the Nemastomataceae rather than the Phyllophoraceae. The most peculiar of all the members of the Phyllophoraceae is *Ahnfeltia plicata*, in which only superficial nemathecia producing monosporangia are known. There is little information as to how these are formed and no details regarding cytology or spore germination. Other species of *Ahnfeltia* (Mikami,

1965) appear to possess somewhat different reproductive structures although the information on these is extremely scanty. One of the most interesting points made by Mikami (1965) concerns the transfer to *Ahnfeltia* of the problematical *Besa gracilis*. The latter is very close to the type species, *B. papillaeformis*, described by Setchell (1912) growing epiphytically or hemiparasitically on a crustose base resembling *Hildenbrandia*. The problems associated with the latter are discussed elsewhere (see p. 202). Apart from the recent work of Newroth, the only data available for the Phyllophoraceae concern the morphology of the reproductive structures. On this the information presented by different authors is frequently inconsistent.

The final topic to be considered relates to the genus *Hildenbrandia*. Attention (Dixon, 1963d) was drawn to the encrusting forms for which only tetrasporangia had been reported and their possible relationship to the life histories of algae of different aspect for which only gametangial and carposporangial phases were known. The value of this suggestion has been amply justified by recent developments (see p. 195). The genus *Hildenbrandia* is an obvious target in this respect although the situation with regard to its various marine and freshwater representatives is still very confused. The systematic position of the genus, which had been referred to both Peyssoneliaceae and Corallinaceae, was disputed for many years until Ogata (1954) proposed that it should be placed in a family of its own, a suggestion supported by the most recent work (Umezaki, 1969). The latter investigation was continued for some time and although the small disc-like crusts formed by the germination of tetraspores grew to be of considerable size they did not themselves form any reproductive structures or give rise directly to any further growth. Tetrasporangia appear to be the only form of reproductive structure reported for the marine species so that the growth of tetraspore germlings to reproductive maturity and the identification of these is obviously an outstanding problem. *H. prototypus* is an organism with an extremely wide distribution both geographically and ecologically and it is possible that several different organisms are being confused. The situation in the Pacific is further complicated by the possible relationship with the incompletely known genus, *Besa*. *B. papillae-formis* was initially described (Setchell, 1912) as small carposporangial papillae growing on a base resembling *Hildenbrandia*, so much so that he considered the possibility that it might represent the gametangial/carposporangial phase of the latter. A second species, *Besa gracilis*, has been transferred (Mikami, 1965) to *Ahnfeltia* (see p. 202), whilst a further additional species, *Besa stipitata*, described recently (Hollenberg and Abbott, 1968) is said to have a crustose base much thinner and somewhat different from that of *Hildenbrandia*. The freshwater *H. rivularis* has been investigated many times and it is known that reproduction by gemmae, stolons, fragmentation, and monospores may occur (Starmach, 1952; Flint, 1955; Nichols, 1965) but there is no evidence for the occurrence of tetrasporangia.

One very curious feature of many of these studies concerns the frequency with which monosporangia, hitherto undetected, were produced on material in culture (Dixon, 1970). Organisms where such monosporangia have been detected in culture but not otherwise reported include *Pseudogloiophloea confusa* (Nemaliales) and *Halymenia floresia*, *Thuretellopsis peggiana* and *Pikea californica* (Cryptonemiales). The monosporangia in these genera are but little differentiated and the process of reproduction appears to consist of little more than the release from a cell of its contents. The cells which behave in this manner frequently develop in pairs, on either side of elongate filaments, and there are some cases where the structures are formed but with no evidence of any release. The most critical aspect of this form of reproduction is that it has been detected only in laboratory culture and there is still no evidence for its occurrence in the field.

To summarise, it is obvious that for the Florideophyceae, there is a mass of information available on various aspects relevant to a consideration of their life histories. In many cases the data available are not sufficient to meet the criteria which were considered necessary for the establishment of a life history (see p. 186). In a few instances, the data are relevant, adequate and well established. On the basis of the latter, there are four situations which appear to be enough to serve as a basis for comparison (Fig. 31); these are:

1. *Polysiphonia:* the life history consists of a sequence of gametangial, carposporangial and tetrasporangial phases, the first and last being morphologically similar, and with the carposporangial phase developing on the gametangial phase.
2. *Bonnemaisonia:* the life history consists of a sequence of gametangial, carposporangial and tetrasporangial phases, all three of which are morphologically dissimilar, and with the carposporangial phase developing on the gametangial phase.
3. *Lemanea:* the life history consists of a sequence of gametangial and carposporangial phases which are morphologically dissimilar, with

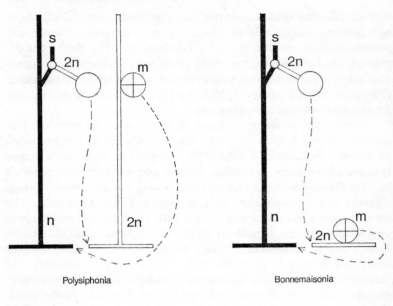

Polysiphonia

Bonnemaisonia

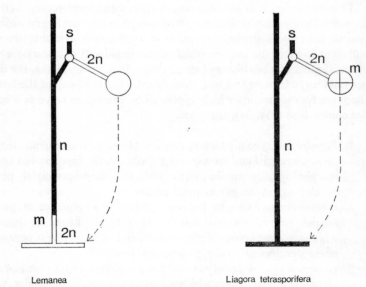

Lemanea

Liagora tetrasporifera

Fig. 31

the carposporangial phase developing on the gametangial phase and with meiosis occurring somatically during the development of the latter.

4. *Liagora tetrasporifera:* the life history consists of a sequence of gametangial and tetrasporangial phases which are morphologically dissimilar, with the tetrasporangial phase developing on the gametangial phase and meiosis occurring in the tetrasporangia.

Life histories in the Bangiophyceae

If one accepts the criteria for the establishment of an algal life history listed previously (see p. 186), *viz.* the occurrence in the field of each morphological phase, the germination of each reproductive body and the growth to maturity of the resulting germling, the counting of chromosomes in each morphological phase and the determination of the positions of syngamy and meiosis, there is not one member of the Bangiophyceae for which the complete life history is known.

The data are most complete for the genera *Porphyra* and *Bangia* but even here they are far from satisfactory. Even the sequence of nuclear phases is still by no means certain in any species. This problem is obviously related to the occurrence or absence of gamete fusion for which the evidence (see p. 174) is more than a little contradictory. There are in fact three different accounts of the sequence of nuclear phases. It is thought by some (Dangeard, 1927; Tseng and Chang, 1954, 1955) that in *Porphyra* meiosis occurs immediately prior to the formation of the so-called 'carpospores' and that this is preceded by syngamy. On the other hand, it has also been claimed in both *Porphyra* (Magne, 1952; Yabu and Tokida, 1963; Kito, 1966) and *Bangia* (Yabu, 1967) that syngamy occurred but that the subsequent divisions were mitotic. Thus, the first group of workers consider that the 'carpospores' are haploid whereas the second group regards them as diploid. Finally, there is the third group of workers who were unable to obtain any cytological indications of either syngamy or meiosis in *Porphyra* (Krishnamurthy, 1959) or *Bangia* (Richardson and Dixon, 1968; Sommerfeld and Nichols, 1970b), with the same chromosome count occurring in all phases of the life history. The evidence (Giraud and Magne, 1968) that meiosis occurs in the fertile cell-rows of the *Conchocelis* phase of *Porphyra* is in agreement with the data presented by the second group of workers.

The occurrence of a *Conchocelis* phase in the life histories of *Porphyra* and *Bangia* is now accepted without question and the original concept that such filamentous growths represent some form of degeneration or aberration induced by the conditions of culture has been completely abandoned. During the past few years there has been a considerable amount of interest in the factors which determine the interconversion between *Conchocelis* and the alternate 'leafy' phase. While the idea that gamete fusion is the necessary trigger, with cells which have fused forming 'carpospores' which germinate to form *Conchocelis* and cells which have not been involved in fusion simply repeating the parental phase through the formation of 'monospores', is superficially attractive, it is difficult to reconcile this view with the discovery that the formation and fate of reproductive bodies is governed by photoperiod in both *Bangia* (Richardson and Dixon, 1968; Richardson, 1970) and *Porphyra* (Dring, 1967; Rentschler, 1967). The so-called 'carpospores' appear to be formed only under 'long-day' conditions, with more than 12 hours light per 24 hours. Once formed, their manner of germination is also controlled by the photoregime to which they are subjected. Under long-day conditions the 'carpospores' germinate in a unipolar manner to give the *Conchocelis* phase whereas under 'short-day' conditions the germination is bipolar and the germling develops into the 'leafy' phase. The various ways by which the *Conchocelis* phase gives rise to reproductive bodies was a problem for many years. The *Conchocelis* phase may form unicellular reproductive bodies occurring singly and these give rise to further *Conchocelis* phase on germination. In addition, elongate cells which become divided into smaller cells through the development of transverse walls are also produced on *Conchocelis*. These are the so-called 'fertile cell-rows' and the spores to which they give rise germinate to form the alternate phase in the life history. The manner of formation of spores by the *Conchocelis* phase would appear to be under strict photoperiodic control both in *Porphyra* and *Bangia*.

The original description of *Conchocelis rosea*, the alga with which the filamentous phases in the life histories in *Porphyra* and *Bangia* have now been identified, was based on material growing in shells from deep water and this has given rise to the unfortunate assumption that this is the only situation in which *Conchocelis* is found. Subsequently, *Conchocelis* has been reported growing in the plates of barnacles and in various other endozoic situations and also in a free-living state on rock substrates. In general, the free-living *Conchocelis* differs somewhat from material growing in shells in that the chloroplasts are more conspicuous, being ribbon-shaped and parietal in position. The appearance of the free-living *Conchocelis* phase of *Bangia fuscopurpurea* is such as to make certain that the filamentous growth obtained in cultures of the freshwater *B. atropurpurea* (Belcher, 1960) and compared with *Rhodochorton* is of this nature.

The idea that thalli of the 'leafy' phases of *Bangia* and *Porphyra* are ephemeral also needs to be revised. The origin of new fronds of *Porphyra* by proliferation from the basal disc after the original frond had eroded away was described many years ago (Grubb, 1923b) although subsequently ignored. Studies of *Bangia* and *Porphyra* on the Pacific coast of North America (Dixon and Richardson, 1969b) have shown that the basal parts can be as persistent as those of the Florideophyceae (see p. 211) and that this form of perennation plays an important role in the biology of these algae. The basal part of the frond as well as the mature frond in both *Porphyra* and *Bangia* may form monospores which on germination give rise once again to the erect 'leafy' phase. These monospores are of the sort described by Drew (1956) as Type II and represent nothing more than vegetative cells, or the contents of vegetative cells, disassociated from the parent thallus. Although reported on several occasions in the literature these monospores were ignored in most considerations of the life histories of *Porphyra* and *Bangia* until the Japanese workers began to investigate their possible significance in relation to the commercial harvesting of *Porphyra*. The frequency with which these reproductive bodies are formed varies considerably, even in the same species. For instance, in *Bangia fuscopurpurea*, the formation of monospores can be detected with relative ease in the British and Californian specimens but not at all in plants from more northerly parts of the Pacific coast of North America (Richardson, 1969).

The general picture of the life histories of *Porphyra* and *Bangia* may be summarised by stating that the 'leafy' and *Conchocelis* phases arise from spores produced by the opposite phase, while each phase may be repeated by spores of a different sort. The type of spore formed is governed by the photoperiodic conditions under which the phase is existing. In addition, the mode of development of a spore (in addition to its formation) may also be influenced by the photoregime to which it is subjected. Although the sequence of morphological phases may be clearly outlined in this way there is still no clear picture of the sequence of cytological phases for the reasons which have been stated previously. Finally, there is also evidence for the occurrence of a phenomenon analogous to the 'direct development' described previously for members of the Florideophyceae (see p. 196). Dangeard (1927, 1931, 1954) argued that the filaments now referred to as the *Conchocelis* phase gave rise to the 'leafy' *Porphyra* phase. His evidence is not conclusive in that the spores which germinate in this way may also give rise to the 'leafy' phase as a result of bipolar germination under different conditions. The first acceptable record of the origin of the 'leafy' *Porphyra* phase from the *Conchocelis* phase is apparently due to Miura (1961) and many more records of this are now available as well as of the direct development of the *Conchocelis* phase from cells of the 'leafy' phase, both in *Bangia* and *Porphyra* (Dixon and Richardson, 1969b; Richardson, 1970).

There is, however, no information on the cytological aspects of these reports of the direct development of the alternate phases in these genera.

The data on the life histories of the freshwater *Boldia erythrosiphon* and the marine *Porphyropsis coccinea* are almost as complete as that for *Porphyra* and *Bangia* in terms of the sequence of morphological phases, although in neither case has any evidence been obtained suggestive of gametogenesis or meiosis. In *Boldia erythrosiphon* (Herndon, 1964; Nichols, 1964b) the thallus is a saccate tubular structure which develops from a basal disc and which gives rise to superficial monosporangia. Monospores are released and may germinate either to form a compact cushion-like structure or a loose system of branched filaments, both of which are usually attached to the substrate. The compact cushion gives rise to the saccate thallus and serves as its discoid attachment. The loose uniseriate filaments may undergo a change in segmentation pattern to give rise to a basal disc or they may produce monosporangia. As with the monospores produced by the upright saccate thallus, the monospores produced by the filamentous phase may either give rise to the cushion-like disc or simply repeat the filamentous phase. As stated previously, there are no indications of either syngamy or meiosis and the same chromosome number has been obtained in all phases in *Boldia*.

The thallus of *Porphyropsis coccinea* resembles that of *Boldia erythrosiphon* in many respects, in that it consists initially of a small saccate structure which arises from a basal cushion although it differs in that the sac does not remain intact as in *Boldia* but splits to give a blade-like expanse at an early stage. The sequence of morphological phases in the life history (Murray *et al.*, 1972) of *Porphyropsis* is also very similar to that of *Boldia* although the methods of spore production differ somewhat. In *Porphyropsis*, monospores are formed in two ways on the blade-like portions of the thallus, either by the unequal division of cells (Type 1 monospores) or by the release of whole cells or cell contents (Type 2 monospores).   Spores formed in either manner are released and germinate to form either a compact cushion-like structure or a loose system of branched filaments.   These do not appear to be anywhere near as distinct in *Porphyropsis* as the equivalent modes of development are said to be in *Boldia* in that the two readily inter-convert and intermediate stages are also formed.   The erect saccate frond arises usually from the cushion-like structure although it can develop from the loose filamentous stage.   The filamentous and cushion-like thalli both give rise to monospores and these can develop into either structure irrespective of origin.   The determination of the mode of development does not appear to be under photoperiodic control and no indications of syngamy or meiosis have been detected.   Nothing is known of the chromosome number in this entity.

In all other members of the Bangiophyceae, it is doubtful if the sequence of morphological phases can even be stated with any certainty. The data available are inadequate and often contradictory but may be summarised as follows.   Recent work on *Rhodochaete* (Magne, 1960) showed that the monospore previously described as being formed laterally from a cell of the thallus is in fact a post-fertilisation product and that its formation is preceded by the fusion of a hitherto unknown smaller spore-like cell with the vegetative cell which ultimately forms the 'monosporangium'.   The smaller spore-like cell is interpreted by Magne as a male gamete and this would appear to be correct in that he showed that nuclear fusion follows the cell fusion.   The subsequent nuclear division is mitotic and there are no indications of the occurrence of meiosis so that it would appear that there is at least one additional morphological phase in the life history, as yet undetected.   Super-ficially, there are certain similarities between the two sizes of 'spore' cells found in *Rhodochaete* and the two sizes of spore formed in *Compsopogon*.   There have been various interpretations of the latter (see pp., 169 174) but no complete explanation.   Germination of the 'mono-spores' produces a plant similar to the parent.   In addition, structures identified (Flint, 1947) as 'antheridia', 'trichogynes' and 'tetraspores' and suggestions that a '*Chantransia*' stage occurred in a species of *Compsopogon* must be discounted as the illustrations are so poor and the phenomena so anomalous.   The other report of tetrasporangia in *Compsopogon* (Das, 1963) must also be treated with caution (see p. 173). This provides no indications of the function of the bodies so identified in the life history.   For species of *Erythrotrichia*, despite the very early descriptions of a sexual reproductive process similar in many respects to one of the methods described for *Porphyra* (see p. 177), there is no recent evidence for such a phenomenon and nothing to indicate its possible significance in the life history. Similar comments must also be made about the very different account of gamete fusion (Heerebout, 1968) which involved the fusion of adjacent cells. Also associated with this account was a report of the occurrence of a filamentous phase identified as a *Conchocelis* phase although nothing was known of the way by which this had arisen or of its significance. The only piece of evidence related to the life history in this genus is that monospores, on germination, reproduce the parental phase.

For various species of *Erythrocladia*, little more is known of the life history than for *Erythrotrichia*. A sexual process has been described, although the evidence is even less convincing than in the latter, while the product of spore germination is indistinguishable from the parent (Nichols and Lissant, 1967).

From this discussion, one can conclude only that there is not one member of the Bangiophyceae for which the complete life history is known. The two critical aspects in need of resolution are the nature of the bodies loosely described as 'spores' and gametes', and the existence of a sexual process in members of this class. It is likely that clarification of these two problems will come only from detailed, highly critical observations, initially of material in culture, with this data then considered with respect to behaviour in the field.

Theoretical and biological life histories

Knowledge of the life history of an alga is of critical importance in any consideration of its biology, ecology or geographical distribution. In all but a very few cases, the establishment of algal life histories has been based entirely on the results of laboratory studies. In certain instances this is unavoidable, but for a macroscopic marine red alga the life history can hardly be defined only on such a basis and some consideration of behaviour in the field is imperative. One of the major consequences of the predominantly laboratory orientation of most studies of algal life histories has been that attention has been restricted to a consideration of reproduction by spores and gametes/zygotes. Such phenomena as vegetative propagation and perennation have been ignored by most investigators and the phenomena excluded from treatments of algal life histories. This is acceptable providing that it is appreciated that the results represent only the 'theoretical' life history.

For a full biological treatment of the life history of an alga it is essential that the data obtained from such laboratory investigations are considered in the light of field behaviour, taking particular account of the two phenomena of perennation and vegetative propagation.

Considering first perennation many red algal thalli are obviously perennial and may be collected at any season of the year. Observation of marked plants of *Gelidium* or *Pterocladia* in Ireland (Dixon, unpublished) shows that some erect fronds can persist for as long as five or even seven years. More important from the present aspect of algal life histories are those instances where only a microscopic portion of the thallus persists from one year to the next. The detection of such a microscopic fragment beneath its covering of debris is virtually impossible until the new erect frond develops at the beginning of the next growing season. The microscopic fragments which persist may be of many different forms. The basal disc of a plant with primary heterotrichy is one of the most common methods of perennation. Another, probably of even greater biological importance, is associated with that phenomenon described previously as 'secondary heterotrichy of axes'. The extensive formation of prostrate axes and their attachment at intervals provides a means for considerable perennation. Because of the apparent disappearance of such thalli during the winter, or other adverse period, such species which persist as minute hidden fragments are frequently regarded as annuals. Attention was first directed to this phenomenon by Knight and Parke (1931) and such species termed 'pseudoperennials'. It would appear that the term 'pseudoannual' would be more appropriate.

Although an individual cell or axis may have a life of only a few months, long-term survival on a clonal basis can occur. Evidence has been obtained indicating that populations of *Pterocladia capillacea* can persist for at least 40 years and there is a strong possibility that certain clones of *Pterosiphonia complanata* have persisted for 130 years (Dixon, 1965). For a plant to persist in this way does not require it to be particularly tough or cartilaginous. The occurrence of dense populations of species of *Porphyra* on the same sand-covered rock from year to year when other, adjacent rocks are devoid of the alga, is often due to dying back and the persistence of few-celled basal fragments beneath the sand.

Considering next vegetative propagation, a detailed account of this phenomenon in many red algae was given by Chemin (1928), one of the few workers to consider this problem. Personal investigations (Dixon,

H

unpublished) indicate that vegetative fragments can become detached, continue to grow and reattach in a great many members of the red algae additional to those mentioned in the literature. The facility with which fragments are detached varies from species to species and the ease with which reattachment can occur varies enormously. The rigid polarity of morphological organisation found in the Florideophyceae facilitates considerably the development of new rhizoidal attachments from the basal stump. Reattachment is not always necessary as considerable growths can occur in loose-lying populations which can develop at a depth greater than the effective wave disturbance. Obviously such populations in the North Atlantic will be disturbed, particularly by the equinoctial gales of autumn, although residual populations can persist at suitable depths and in suitably protected places throughout the year. The problem with vegetative propagation is not to demonstrate that it can occur, something which can be achieved under laboratory conditions with practically every species which is examined experimentally, but rather to demonstrate its actual occurrence in the field. This requires considerable time and trouble together with a detailed knowledge of the particular entity under consideration.

The occurrence of either perennation or vegetative propagation, or both, can modify considerably the occurrence of phases of a life history in the field, particularly if one phase or the other is selectively influenced. There are many reports in the literature of species which are said to possess a *Polysiphonia* type of life history, but where one phase is of much higher frequency than the other or has a wider geographical distribution. In Europe, for instance, plants bearing tetrasporangia are usually reported from further north than plants with gametangia, while at the northern limits of distribution most species are totally sterile. Observations of this kind immediately raise questions as to the life history of the alga in the area where only one type of reproductive structure occurs or where all material is completely sterile. Investigation of such a situation demands a study, both in the field and the laboratory, of reproduction and its control, spore production, viability and longevity, perennation and vegetative propagation, using both cytological and cultural techniques. At the present time, the necessary data are either non-existent or unreliable.

Despite this, some progress is being made in a few cases. In the case of *Bornetia secundiflora*, the indications are that a high proportion of thalli perennate from year to year as basal fragments. The population in the English Channel is made up of perennating sterile thalli to which a small number of additions is made each year by vegetative propagation (Dixon, 1965).

One of the consequences of the great interest in the occurrence, distribution and life histories in the Bonnemaisoniaceae has been that a considerable amount of information relevant to the present discussion has been obtained in various species of this family. *Asparagopsis armata* was first reported in the British Isles, together with what is now recognised as its tetrasporangial *Falkenbergia* phase, about 30 years ago. The production of tetrasporangia has been recorded only very rarely (McLachlan, 1967) in British material. The plants of *Asparagopsis* first detected bore carposporophytes which contained carpospores, although in recent years the frequency with which carposporangia are formed appears to have diminished considerably (De Valéra and Folan, 1964). The occurrence of vegetative propagation of the *Falkenbergia* phase was demonstrated by Chihara (1962) and by Dixon (1965), while in the gametangial *Asparagopsis* phase specialised branches with reflexed spines, which become entangled in other algae, are present. With one exception, there does not appear to be a record of *Asparagopsis armata* growing other than entangled in other algae, at least in the British Isles, with no trace of a basal attachment structure which would indicate origin from a spore, so that for *Asparagopsis armata* the sequence of phases characteristic of its life history appears to have been replaced, if not entirely, then at least to a very considerable extent by independent vegetative propagation of the gametangial and tetrasporangial phases (Dixon, 1965; Martin, 1969).

Discrepancies such as these, between the actual biological life history of the species demonstrated in the field and what has been accepted as the 'theoretical' life history, also raise a further important question. It is usually assumed that the sequence of events demonstrated in a laboratory investigation represents the total expression of the life history. It must be appreciated that there may well be several expressions of life history possible for any species depending upon conditions and that to assume on 'theoretical' grounds the existence of a single model is never justified. In conclusion, one can only reiterate the crucial need always to integrate the results of laboratory experimentation with field data.

5

Economic Utilisation

The earliest record of seaweed utilisation appears to be in a Chinese herbal of the third century B.C. and it is known that algae have been used as a food source since that time.   The extensive use of brown algae as fertilisers was established in Europe by the twelfth century and apparently in the Orient well before that date.   There was extensive burning of kelp, first for the production of alkali in the eighteenth and subsequently for the manufacture of iodine in the nineteenth century. The present century marks the establishment of the real industrial use of algae.   The number of genera and species of algae which it is known are, or have been, used as an item of diet or from which some material has been extracted is enormous (Levring *et al.*, 1969).   Some instances represent industrial activities of major economic importance whereas others are merely reports of some particular or uncommon material extracted once from a marine alga.   In terms of the red algae, the two major uses are for human food and to serve as a raw material to be extracted.   The latter is by far the more important economically. There are a few reports of the use of marine red algae as animal fodder, but these are relatively minor compared with the extensive use of brown algae for that purpose.   The latter are also used widely as an organic soil dressing for which the red algae appear never to have been used, although deposits of coralline algae are used extensively along the coasts of Europe as a source of lime for soil dressing.

Red algae as animal and human food

Although the use of algae as a staple item of human diet has been extensive for several centuries in China, Japan, the Indo-Malayan region, Hawaii and the Philippines, algae have never served in other areas as more than a food additive and, in recent years, even this

214

form of utilisation has declined considerably (Newton, 1951; Chapman, 1970).

In western Europe and on the Atlantic coast of North America, the two most important red algae used as food are *Rhodymenia palmata* and various species of *Porphyra*. The former has been known by a variety of colloquial names ('dulse', 'dillisk', 'crannough', etc.) in Scotland and Ireland. The alga was either eaten raw, in a fresh or dried condition, or cooked like spinach. There is still a certain amount of local collecting for personal use, but the extent of utilisation has declined in Scotland and Ireland to such an extent that no commercial trade exists at the present time. There has been a similar decline during the present century in the use of *R. palmata* as human food in Iceland, where there has been an active commercial trade in the dried product since the twelfth century. At the present time, the only commercial use of *Rhodymenia* in Iceland appears to be as a dietary supplement in the stall feeding of cattle during the winter. Despite the decline in the commercial utilisation of *Rhodymenia palmata* in Europe, there is still a very active trade in this product in the Canadian maritime provinces (MacFarlane, 1964). The quantity of dried *Rhodymenia* marketed each year is of the order of 40-50 tons. The plants are harvested during the summer months and dried as quickly as possible to avoid decay. Initially the dried plant is black but returns to a more reddish colour, probably due to the slow re-absorption of moisture. The product is sold in this condition and eaten raw in the dried state. Bowls of dried *Rhodymenia* are to be found on tavern counters and the principal use of the product appears to be for the induction of thirst. The drying of the *Rhodymenia* is undertaken at any suitable location and the material can be seen, spread out to dry on waste ground, on roadsides, and even on the road itself on occasions. However, the product cannot be sold other than in a polyethylene bag because this would not be hygienic!

Species of *Porphyra* have been used extensively for human consumption in Ireland and Wales and there is still a considerable trade in this commodity in South Wales where the consumption is of the order of 200 tons (wet weight) per annum. The interesting feature of this industry is that the raw material is almost all imported from coastal localities outside the principal areas of South Wales in which it is eaten, with only a minute proportion collected locally and processed individually (Hampson, 1957). The product is known locally as 'laverbread' (*bara lawr*, in Welsh) and the commercial trade is now highly organised with bulk importation, processing and large-scale distribution for market-day and weekend sales. Detailed analyses of the weed origin (Hampson, 1957) show that the supply in 1954 was maintained by permanent collectors from certain regions of England (e.g. Cornwall and Cumberland), Scotland (e.g. Dunbar and the Solway Firth) and North Wales, with small quantities imported from various localities in Ireland when supplies from more accessible collecting grounds failed. The processing commences with several washes in freshwater to remove sand and debris, after which the weed is boiled for several hours in 60-gallon open-topped boilers with salt added at the rate of about 3 lb per 100 lb of weed. When the fronds have been reduced almost to a pulp they are removed from the boiler, allowed to drain and then minced; the product is marketed and sold in this condition. A certain amount is even shipped to markets in parts of England, although the quantity involved is small. It is always prepared for eating by warming in fat and it is sometimes made into cakes and coated with oatmeal.

In Ireland, a considerable quantity of *Porphyra* is collected for personal consumption as well as for export. The local method of preparation involves either frying with fat or, alternatively, converting to a pinkish jelly by heating the fronds in a saucepan with a minimal quantity of water and beating with a fork. Considerable quantities of *Porphyra* were exported from Ireland at the beginning of the present century (Cotton, 1912) and although the amount exported is much reduced, it is not non-existent as is sometimes claimed (Chapman, 1970). About 20 tons (wet weight) were exported in 1954, of which about one third went to South Wales. It is probable that the quantity involved here would be greater were it not for the very poor communications between the source of the product and the area of demand. The remaining quantity exported from Ireland is sold in those areas of England (Bristol, Birmingham, Coventry, etc.) with a high Irish immigrant population. It is difficult to ascertain the precise details of the trade in that the total quantities involved are small and the negotiations between supplier and vendor are often arranged on a personal, family basis.

In addition to *Rhodymenia palmata* and the various species of *Porphyra* used in Ireland, Scotland and Wales, there are many species of red algae for which there was some minimal use. About twenty species of red algae had local Erse names, indicating some degree of familiarity if not active utilisation. Of these *Laurencia pinnatifida* was probably the most widely used, as a condiment, known colloquially as 'pepper dulse'. It is said by various authorities that *Dilsea carnosa* (previously known as *D. edulis*) was used as a food species in various parts of the western seaboard of Europe. It is said that this species, if prepared by pinching between two hot irons tasted better than *Rhodymenia palmata* (Sauvageau, 1920). While this comparative statement might well be accurate, the general texture of *Dilsea* resembles most closely a heavy-duty inner tube and the claims of its use most likely represent an old taxonomic misidentification of the species of laminate red alga which was widely eaten rather than an accurate statement of edible quality.

Japan is the most important producer and user of algae as human food. Of the various species of commercial importance, species of *Porphyra*, known collectively as *nori*, are probably the most important of the red algae. Various species are collected from 'wild' populations although the major commercial production is by cultivation. *P. tenera* is probably the most widely used species although several others are involved also. Artificial cultivation of *Porphyra* species was first undertaken in the seventeenth century in Tokyo Bay and is now of considerable importance in any area of shallow water. Bundles of bamboo or brushwood are sunk into the mud in water of depth from 3 to 5 m; in most cases, a combination of bamboo and netting is used to produce a raft-like structure covering an area of about 40×3 m. For many years, the yields of *Porphyra* fluctuated enormously prior to the detection of the *Conchocelis* phase in the life history of *Porphyra* in 1949 (see p. 205) and the appreciation of the significance of this hitherto undetected phase. Previously, the yield of *Porphyra* depended on the production of spores by the highly erratic 'wild' *Conchocelis* phase. The fluctuations in settlement have been overcome by artificial cultivation of the *Conchocelis* phase. Shells containing the *Conchocelis* phase are attached to the rafts or the nets are dipped in baths to which crushed shell has been added, in order to increase the level of inoculation. The *Conchocelis* phase liberates the spores from which the *Porphyra* will develop between September and December, and by January the beginning of the *Porphyra* is ready to harvest. In some cases, the nets or the superficial brushwood, with its attached young *Porphyra* plants, are taken to another locality for the final growth. The conditions which determine settlement and growth are somewhat different in that high salinity is preferable for settlement whereas maximum growth occurs in areas of lower salinity in which nitrogenous material is abundant, i.e. in the vicinity of sewage outfalls. Although the use of the artificially-prepared *Conchocelis* helped to increase output considerably and stabilise productivity in the early 1950s, there are now serious problems developing as a result of industrial pollution of the shallow-water areas in which *Porphyra* cultivation is carried out. It has been suggested, and the experimental work is now under way, that following the highly successful artificial cultivation of the *Conchocelis* phase there was no reason why the entire process could not be conducted under artificial conditions. When the plants of *Porphyra* are collected, they are washed in freshwater in order to remove sand and mud and then finely chopped. The chopped material is then spread on frames to dry. When dry the thin sheets of dried chopped *Porphyra* are stripped from the frames and marketed. The dried sheet of *Porphyra* may be toasted and sprinkled on the morning rice. Flavoured boiled rice and strips of seafood, usually eel or shrimp, egg or certain vegetables may be rolled in a *Porphyra* sheet as *sushi*. There was a steady increase in the productivity of *Porphyra*, from a total production of 24 000 tons (wet weight) in 1927 to 130 000 tons (wet weight) in 1961 although in the past decade productivity has been stationary or even declined slightly as a result of the pollution problems mentioned previously.

There are various other red seaweeds which are collected in a 'wild' state and eaten by the Japanese either in the fresh state or dried, and prepared in a wide variety of appetising ways. In Indonesia and the Philippines, many marine algae are eaten raw or after a brief immersion in boiling water, either with a spicy sauce or a sugar sauce. On the mainland of Asia, edible seaweeds have been articles of commerce in China for many centuries although the bulk of the seaweed eaten was imported when the most recent figures were available (Kirby, 1953). Nothing is known of the present state of affairs, although from the scientific literature there are many obvious indications that serious attempts are being made to develop a native edible seaweed industry, based principally on the Phaeophyta. By comparison, little use is made of edible red seaweeds in either India or Burma, the two genera most used in the latter (*Catenella, Bostrychia*) not being used anywhere else in the world as far as can be ascertained. Large quantities of seaweed were eaten in past times throughout Polynesia, particularly in Hawaii. About a century ago, over 50 species were eaten of which some 30 red algae were eaten extensively. The usual practices were to mince the algae and use them in the raw state in combination with other foods or to use them as a vegetable. Seaweed eating has now declined considerably. Surprisingly little seaweed eating took place in Australia while in New Zealand the native population used few seaweeds of which *Porphyra* was probably the most frequently eaten. One of the curious features of the algae eaten in various parts of the world was that they tend to be restricted to a relatively few genera.

In addition to those red seaweeds which were eaten as a whole, either in a raw or cooked condition, various marine red algae have been used by man over the centuries as a source of gelatinous material. It is difficult to draw the line between such personal extraction procedures and the large-scale industrial extraction of agar, carrageen, etc., which is discussed in detail in the following section. Most of the latter were, in fact, based on prior reports of such personal extraction. One can justify distinguishing between the uses of *Porphyra* and *Chondrus crispus* in that the *Porphyra* as a whole is eaten whereas that portion of the *Chondrus* which is used is only a derivative product. The use of *Chondrus crispus* as a source of an edible gelatinous product is of considerable antiquity in Ireland and Scotland and it would appear to have been introduced into North America by the early years of the nineteenth century. The weed after collection is washed to remove sand and then spread out, usually on a convenient patch of grass, to bleach and for the salt content to be reduced. After drying the weed is ready to use. In most instances it is added to milk and the gelatinous carrageen content extracted by gentle heating after which the bunch of algae is removed. The resultant blancmange-like product is insipid to most palates unless sugar, lemon rind or other flavouring has been added. Similar extracts are obtained from various species of red algae in different parts of the world. Species of *Eucheuma* are employed extensively for this purpose in the East Indies and Philippines while carrageen-like products are obtained on a personal basis from species of *Hypnea*, *Iridaea*, *Gracilaria* and *Grateloupia* throughout the Pacific region. The use of *Suhria vittata* by the early colonists in South Africa for the same purpose is discussed in detail by Chapman (1970).

Commercial extraction from red algae

The term phycocolloid was first used by Tseng (1946) to refer to the polysaccharide complexes extracted from red and brown algae which form colloidal systems when dispersed in water. The characteristic polysaccharides of the red algae are mucilages which contain varying proportions of D- and L-galactose, 3,6-anhydro-D- and L-galactose, monomethylgalactoses and ester sulphates (Percival and McDowell,

1967). A bewildering number of polymers based on the above units has now been isolated and characterised from various genera of Rhodophyta although the commercial products are frequently considered as falling into two principal classes, agar-polysaccharides and carrageenan-type polysaccharides. A third category of polysaccharide, the porphyran-polysaccharides, obtained from members of the Bangiophyceae, has no known use as an extracted material although it obviously must compose a major nutritional component of the *Porphyra* which is consumed as human food (see p. 215).

Agar

Agar is probably the best known of all algal products. The name is of Malaysian origin and was probably applied originally to the extract of a species of *Eucheuma* although, as will be seen, a distinction must be made between the product extracted from representatives of that genus and the substance to which the name agar is now applied. The pharmaceutical definition of agar cites a phycocolloid of red algal origin which is insoluble in cold but readily soluble in hot water with a 1·5% solution being clear and forming a solid and elastic gel on cooling to 32-39°C, not dissolving again at a temperature below 85°C. In terms of its chemical structure the principal properties of agar are often said to be a consequence of its being a mixture of two principal components, agarose and agaropectin, with the former providing the characteristic properties. Agarose was considered to be a neutral molecule whereas agaropectin was charged. This view requires some modification in view of the recent results of Duckworth and Yaphe (1971a, b), which have shown that agar is a complex mixture of polysaccharides all having the same backbone structure but substituted to a variable degree with charged groups.

Agar is obtained commercially from species of *Gelidium* and *Pterocladia* as well as from various other algae, such as *Acanthopeltis*, *Ahnfeltia* and *Gracilaria*. These algae are often referred to loosely as agarophytes. Commercial production of agar was a world monopoly of the Japanese for many years and even in 1939 that country was still the major producer, despite the development of agar industries in California and elsewhere. Wartime demands in areas deprived of Japanese agar led to the development of agar industries in many countries, some of which have continued and prospered while others have since declined or disappeared. In addition, post-war developments of agar industries have taken place in many additional areas, such as Morocco, where the raw materials are available. The traditional method of preparation utilised extraction by boiling for several hours. Prior to this, the raw material had to be cleaned and bleached in the sun, with several washings in freshwater used to facilitate bleaching. The boiled extract was acidified and concentrated and then cleaned of waste material by repeated freezing and thawing after which it was dried and marketed in the form of cakes or flakes. The more modern process extracts the agar under pressure in autoclaves. The agar is decolorised and deodorised by activated charcoal, filtered under pressure and evaporated under reduced pressure. Further purification by freezing is then undertaken. The quality of the product varies considerably, depending upon the species from which it has been obtained, the collection, drying and bleaching procedures, and the manufacturing techniques. The features upon which quality is estimated are the strength of the gel, the temperatures for melting and gelling, the colour and the ash content. It has been shown (Cooper and Johnstone, 1944) that the maximum yield of agar is obtained in *Gelidium robustum* (as *G. cartilagineum*) from plants harvested during the summer whereas for *G. amansii* it was found (Asano *et al.*, 1951) that the agar content rises during the summer with a maximum in the autumn and a decline through the winter. The indications are that each species of *Gelidium* has a slightly different behavioural pattern with respect to the gel content.

The bulk of the agar used today is obtained from Japan, Korea and Morocco. Smaller quantities are produced in many countries in Europe (France, Italy, Spain), Africa (Egypt, South Africa), Asia (Ceylon, India, Indonesia) and the Americas (California, Mexico, Brazil, Chile). Several surveys of the production and uses of agar have been published in recent years (Humm, 1951; Boney, 1965; Chapman, 1970). The most important features governing the demand for agar is the considerable strength of the gel, even at extremely low (1%) concentrations. Agar is used in various industries where material with minimal strength requires stiffening prior to cutting. The greatest use of agar is in association with food preparation and technology and with the pharmaceutical trade. Agar is used for gelling and thickening purposes, particularly in the canning of fish and meat whereby the undesirable effects of the can are reduced and there is some protection against shaking of the product in transit. It is used in the manufacture of processed cheese, mayonnaise, puddings, creams and jellies. Pharmaceutically, agar is used directly as a laxative, but more frequently as an inert carrier for drug products where slow release is required, as a stabiliser for emulsions, and as a constituent of cosmetic skin preparations, ointments and lotions. The use of agar in bacteriology and mycology, as a stiffening agent for growth media, for which this material was first used almost a century ago, is still a very considerable part of the total demand.

Carrageenan

Carrageenan is the extract of 'carrageen', or 'Irish Moss', the trade names for a mixture of *Chondrus crispus* and various species of *Gigartina*, particularly *G. stellata*. As will be seen, carrageenan is a mixture of materials whose proportions vary depending upon the species from which it has been extracted as well as upon the season of collection. Precise definition of carrageenan is still not possible, although two major components have been isolated. These are a branched chain polymer of the kappa fraction and a linear chain of the lambda fraction. Kappa-carrageenan has a molecule with a similar backbone to agarose (see p. 220) except that it contains anhydro-D-galactose in place of the L-isomer while D-galactose units of agar are sulphate esters in carrageenan. Lambda carrageenan is a linear D-galactose sulphate. These two fractions can be separated by physical means although the development of enzymes specific to a particular type of linkage has assisted considerably in the interpretation of chemical structure and the quantitative evaluation of the occurrence of the different fractions (Yaphe, 1959).

Commercial extraction is similar to that for agar, although unlike agar the carbohydrate polymer cannot be purified by freezing. The dried Irish Moss is washed with freshwater to reduce the salt content and then boiled for some hours with two to four parts of alga to 100 parts of water. The soluble carrageenan is separated from the insoluble residue in a centrifuge. Following filtration and some evaporation under vacuum, the carrageenan is dried on a rotary drier. As with agar, there are considerable seasonal and environmental variations in gel strength as well as quantitative changes in the amount of carrageenan which can be extracted. In *Chondrus crispus*, minimal polysaccharide content occurs in the spring and early summer with the maximum in late summer and autumn (Butler, 1936) and an analysis of variation in gel strength (Marshall *et al.*, 1949) showed that it also improved from spring to autumn. In addition, with *Gigartina stellata*, variations in gel strength occurred in relation to harvesting procedures (Marshall *et al.*, 1949). Yield and gelling strength were low during April in areas from which material had been cropped during the previous year although extracts from plants not touched in the previous year were higher. Subsequent collections showed an increase in quantity and gel strength with a maximum in September followed by a slow decline during the winter.

The major production of carrageenan from *Chondrus crispus* and *Gigartina stellata* takes place in Canada and the United States, followed by France, Ireland and Norway. The terms agar and carrageenan were very confused initially and although some clarification is appearing some of the literature requires careful analysis before the term accepted for the mucilage under discussion is accepted. Although at one time, the product of *Chondrus* was frequently referred to as agar, this is now not normally the case, other than in more popular and general literature. The difficulty in the application of the terms was due to the inability to define the materials agar and carrageenan in strict chemical terms, the frequent impossibility of identifying the species from which the extract was obtained or to indicate its geographical place of origin, and the many occasions on which a commercial extraction was made of a mixture of species from a wide range of localities. The major chemical identification is now complete although only in terms of the two or three major components of the most widely used products. There are now many additional species and genera being utilised as sources of polysaccharides which are referable to carrageenan. Such gels are softer than that of a true agar and they are frequently referred to as agaroids. Species of *Phyllophora* and *Eucheuma* are used extensively for this purpose, while in Denmark a carrageenan, more specifically furcellaran, is obtained from the floating populations of *Furcellaria fastigiata* in the Kattegat.

Carrageenans are used extensively for many of the same purposes as agar. Because of the lower gel strength, carrageenans are used less for stiffening purposes than agar, although for stabilisation of emulsions in connection with paints, cosmetics and other pharmaceutical preparations, carrageenans are preferred to agar. For the stiffening of milk and dairy products, such as ice-cream, carrageenans have supplanted agar completely in recent years and it is in this area that demands for these products are greatest. One particular use is in connection with 'instant' puddings, sauces and creams made possible by the use of this stabiliser which does not require refrigeration.

Gelans

In addition to the more frequent polysaccharide mucilages there is a great number of substances which, from the meagre information available, resemble kappa-carrageenan except that the degree of sulphation is much less. The detailed chemical structure and characteristics of most of these substances have not been elucidated although a number are known in general terms (O'Colla, 1962).

The use of coralline algae for soil dressing

Several species of crustose coralline algae can occur in an unattached 'loose-lying' state, producing extensive beds in the shallow sublittoral. Such beds occur in various parts of the Mediterranean and along the western coasts of Europe at depths ranging from 6 to 75 feet below extreme low water. These beds are composed principally of two species, *Phymatolithon* (*Lithothamnium*) *calcareum* and *Lithothamnium corallioides* (Cabioch, 1969; Adey and McKibbin, 1970), which are known locally under the collective names 'marl' in England, and 'maerl' in France. Loose-lying plants are frequently cast up from shallow water in quantity in the equinoctial gales as are these coralline algae, despite the calcareous incrustation and their higher density. As a result of this, the local 'sand' is composed entirely of partially or complete crushed thalli of coralline algae. In the living state the thalli are rose-pink but become bleached white on being cast up. Deposits of such calcareous material on the beaches may be recognised from a distance because of their dazzling white appearance. Such beaches are usually referred to as 'coral beaches' in western Ireland (Cotton, 1912) and southwest England and their frequency is indicated by the number of bays named 'Whitesand' or 'Whitestrand'. Some of the beaches may contain a proportion of crushed shells in addition to the calcareous algae.

The coralline sand is carted for use in liming the highly acidic soils which are found along much of the European coast. In Ireland, the rate at which new material is cast up is adequate to replenish the quantity removed. The use of 'coral sand' in England, Ireland, and Wales appears to have diminished steadily during the present century due partly to the abandonment of the poorer, marginal, coastal land and partly to improved transportation facilities and the reduced costs of other sources of lime. In northwest Spain, by comparison, the use of calcareous sand, of both shell and coralline algal origin, appears to be still increasing, particularly in Galicia. The most extensive use of coralline algae occurs in Brittany. Extensive sublittoral beds occur off the southern coast although on the northern coast they tend to be restricted to the submerged valley bottoms (Lemoine, 1910; Lami, 1941). The calcareous material is collected partly as sand from the beaches but the greatest quantity is obtained directly from the sublittoral beds by grab or dredge. The rate of utilisation is so high that there have been suggestions of exhaustion, particularly around Concarneau.

Analyses of the calcareous material vary considerably. Obviously, the extent to which the deposits of calcareous algae are contaminated with shell fragments is the source of some of this variation. Even when relatively pure deposits of coralline algae are analysed there is still considerable variation in chemical composition, due partly to the source of the material. Material collected directly from the sublittoral, as in Brittany, is the most superior fertiliser. It contains a certain quantity of nitrogenous and phosphoric substances although these do show some seasonal variation. For the beach material, the variation depends principally on the length of time the thalli have lain on the beach and the extent to which bleaching and crushing have occurred. The highly bleached and crushed beach material consists of little more than the calcite skeleton of the alga. In Galicia, it is recommended that August and September are the best months for collection, even of beach material. At these times a considerable enrichment of nitrogen occurs, although this is said to be due to crustacean infestation (López-Benito, 1963).

6

Fossil Rhodophyta

As with most groups of algae, the fossil record for the Rhodophyta is poor. Phycologists who rarely pay attention to fossil material must be held responsible for some of this ignorance and most of the investigation of the red algal fossils has been undertaken by geologists rather than by specialists in the algae. It is symptomatic that fossil Rhodophyta are virtually ignored in the *Die Gattungen der Rhodophyceen* of Kylin (1956). Most fossil red algae are calcareous fossils interpreted as being referable to the red algae. As the calcareous forms are by far the better known they will be treated first.

Calcareous fossil Rhodophyta

A detailed treatment of the fossil calcareous red algae has been given by Johnson (1961).

Corallinaceae

Most of the calcareous fossils are referable to the Corallinaceae. These occur back to the Jurassic with their most extensive distribution being from the Cretaceous to the present. The degree of fossilisation is excellent in many of the fossil Corallinaceae so that a considerable amount of detailed anatomical and morphological data is available. Unfortunately the specimens are often somewhat fragmentary so that generic or specific assignment of material is difficult despite its excellent preservation. Representatives of both crustose and erect thalli occur as fossils and in only a very few cases are genera known only in the fossil or only in the extant state. Most occur both as living genera and as fossils, with a continuous record from the Cretaceous to the present. As the morphology of the extant forms has been discussed in detail elsewhere (pp. 57, 72) no further comments are necessary.

The extent of the deposits of these fossil Corallinaceae is often considerable, so that the Miocene Leithakalk of Austria is used extensively as building stone. Probably the most significant fossil deposits are those which occur in tropical seas as coral reefs. The name for these calcareous masses is, strictly speaking, a misnomer in that the true animal corals make a much smaller contribution to their formation than do the coralline algae which they contain. The depth of such coral reefs often exceeds 1000 feet, that at Eniwetok Atoll being 4500 feet deep and representing deposits from early Tertiary to Recent. The discovery that many of the earth's oil deposits are associated with ancient coral reefs has stimulated an interest in fossil calcareous algae far in excess of what they would normally receive. As a consequence, although a considerable number of publications has appeared in recent years much of the data is not readily available because of either deliberate or accidental restrictions in circulation of the reports concerned.

Solenoporaceae

The family Solenoporaceae (Johnson, 1960, 1961) is closely related to the Corallinaceae. The genus *Solenopora* and its allies were not separated from the latter family until the early part of the present century and this separation is still not accepted by some. Genera of the Solenoporaceae occur from the Upper Cambrian to the Cretaceous and it is interesting that their decline occurred at the same time as the rapid increase in the Corallinaceae.

The anatomical structure of the thalli in *Solenopora* and the genera of the Solenoporaceae is very similar to that in the Corallinaceae. There are certain differences between the two families, the most significant of which are:

1. The cells in thalli of the Solenoporaceae are much larger, particularly much longer, than those of the Corallinaceae. Also, they tend to be polygonal in section whereas cells of the Corallinaceae are rectangular, although both may be circular on rare occasions.
2. Transverse cell walls are poorly developed in thalli of the Solenoporaceae and may even be absent, but it is not clear whether this is due to weak calcification and poor preservation rather than original absence.

In addition, it has been argued (Johnson, 1960, 1961) that the differentiation into hypothallus and perithallus found in the Corallinaceae is not found in thalli of the Solenoporaceae. These terms have been applied in the Corallinaceae in two very different ways and this situation needs to be resolved before any further use is made of this feature for systematic purposes. The relationship between the Solenoporaceae and Corallinaceae is further complicated by the meagre knowledge of reproductive structures of the Solenoporaceae. A few specimens, from the Silurian, have structures which suggest either isolated sporangia or even small conceptacles, but there is little agreement as to the interpretation of these.

Most of the thalli assigned to the Solenoporaceae occur as rounded nodular masses although a few encrusting forms are also known. It is interesting that throughout the Palaeozoic, the numbers of the Solenoporaceae are low, so that in Silurian and Devonian limestone reefs the animal corals are by far the more important. This is in marked contrast to the situation in more recent reefs where the Coralline algae are much more important than the animal corals (see p. 227). It is usually considered that the Solenoporaceae lived under conditions similar to those where Corallinaceae occur today so that the decline in numbers of the Solenoporaceae during the Cretaceous with the associated appearance and increase in true coralline algae, such as *Sporolithon* and *Lithothamnium*, may well be due to competition between the two groups of calcareous red algae. It is usually assumed that the Corallinaceae evolved from the Solenoporaceae and although the evidence for this is not particularly complete, it is probably better than for any other flights of phylogenetic fancy in the Rhodophyta.

Gymnocodiaceae

The genus *Gymnocodium* was based on the previously described fossil alga *Gyroporella bellerophontis*, which had been interpreted as a member of the Chlorophyta. The generic name indicates that this alga was then regarded as a member of the Codiaceae rather than the Dasycladaceae, before its final transfer to the Rhodophyta (Pia, 1937). This author considered that *Gymnocodium* resembled the recent genus *Galaxaura* and placed *Gymnocodium* in the Chaetangiaceae. Subsequently, a more detailed comparison of larger collections of *Gymnocodium* with *Galaxaura* indicated that although there were various features in common between these two genera, the differences were sufficient for *Gymnocodium* to be placed, not in the Chaetangiaceae, but in a distinct family, the Gymnocodiaceae (Elliot, 1955). This decision appears to be fully justified but it is difficult to see why the Gymno-codiaceae should be placed in the Cryptonemiales (Johnson, 1961).

The thalli of *Gymnocodium* consist of hollow calcareous segments which are cylindrical, oval or cone-shaped, and circular or oval in cross-section. Each thallus is composed of many segments and is rarely branched; the thalli grow upward in a cluster from the point of attachment. The genus *Gymnocodium* is known from the Permian while the very similar and closely related *Permocalculus* has been reported from the Permian to the Cretaceous.

Non-calcified fossil Rhodophyta

Several fossil algae were discussed by the elder Agardh (1822) although their attribution is exceedingly doubtful in view of the meagre descriptions and the absence of illustrations. Subsequent studies by Brongniart (1829) and Schimper (1869) added to the number of described entities while summaries were provided by Hauptfleisch (1897) and Pia (1927). Although many of the entities described are undoubtedly based on algal material, it is obvious that the attributions of some are questionable. The initial authors treated most of the fossils as species of *Fucoides* because of the supposed resemblance to the genus *Fucus* which was the name then used for the heavier, more cartilaginous thalli. Subsequent authors described genera under such names as *Delesserites* and *Polysiphonides* because of their supposed resemblance to the extant *Delesseria* and *Polysiphonia*. The degree of preservation in these non-calcareous fossils is very much poorer than for the calcified thalli so that the level of information on their anatomical detail is very low. Because identification demands a knowledge of the detailed morphology of the thallus and reproductive structures many of the supposed determinations can only be regarded as highly speculative. In most cases, identifications are based solely on the basis of the comparison of external form. In general it can be stated that there are very few non-calcareous fossils which can be referred with certainty to the red algae while knowledge of their detailed structure is virtually non-existent.

The recent increase of interest in post-glacial chronology and archaeology is likely to provide 'fossil' material of more recent age to that discussed previously. To date, little algal material other than diatoms has been detected, but it may be that this is simply a consequence of failure by archaeologists to recognise dried fragments of red algae. A recent study (Illman *et al.*, 1970) of a deposit from the 'Champlain

Sea', dated at 10 800 ± 150 years old, disclosed a flat leafy thallus, probably of *Rhodymenia*, endozoic *Audouinella membranacea* in a hydroid in the deposit and material of a species of *Acrochaetium*, in addition to a number of fragments of macroscopic algae of other groups.   Archaeologically, one of the few cases where red algal material has been identified is a pre-columbian deposit in Peru, dated at the fourteenth century, where *Gigartina chauvinii* was identified.   This discovery is of particular significance in that this species is still sold in Peru as food for human consumption (Howe, 1914).

7

Systematics and Phylogeny

It is not proposed to give a detailed treatment of the systematics of the red algae but rather to outline the way by which the currently accepted system developed and to indicate its strengths and weaknesses. Those interested in the minutiae are referred to Kylin's *Die Gattungen der Rhodophyceen* (Kylin, 1956). This critique will be followed by a brief indication of the principles and procedures by which future advances are likely to occur. Finally, an attempt will be made to outline the phylogenetic relationships of the Rhodophyta and to give an outline of the Orders, Families and Genera with reference to those mentioned in the present text.

Systematics of the Rhodophyta

Historical introduction

The nomenclature of red algae, as with that of most groups of plants, begins with the *Species Plantarum* of Linnaeus (1753). The fundamental criteria used in that work related to the distribution and arrangement of phanerogamic reproductive structures. As a result, the phanerogams were distributed into 23 Classes while the algae, together with the fungi, mosses and pteridophytes, were lumped together into the 24th Class, the Cryptogamia, so named because phanerogamic reproductive structures were not displayed. Linnaeus accepted 14 genera of algae of which four, *Conferva*, *Ulva*, *Fucus* and *Chara*, contained only organisms now regarded as algae. In addition, two other genera, *Byssus* and *Tremella*, contained a few species of algae although most were fungi. Species of red algae treated by Linnaeus were referred to the three genera *Conferva*, *Ulva* and *Fucus*. The first received the filamentous species or those with particularly slender, elongate thalli, while species with flat membranaceous thalli were

231

referred to *Ulva* and those with fleshy or cartilaginous thalli were placed in *Fucus*. The Linnaean circumscription of these three genera survived for almost half a century although the number of described species referred to them increased very considerably during this time. As a consequence of the great increase in number of species the artificiality of the Linnaean genera was appreciated increasingly so that during the last few years of the eighteenth and first three decades of the nineteenth centuries considerable numbers of new genera were described. Although Stackhouse (1795-1801, 1809, 1816) was probably the first to suggest breaking down the three Linnaean genera, his work was largely ignored for various reasons whereas the slightly later proposals of Lamouroux (1813) and the elder Agardh (1817, 1820, 1822, 1828) were generally accepted, forming the basis for the currently accepted generic taxonomy not only of the red but also of the other major groups of algae as well.

Lamouroux (1813) was the first to segregate on the basis of colour certain algae now placed in the Rhodophyta from others of similar external morphology. For these he established the category Floridées from which the name of the currently accepted major class of red algae, the Florideophyceae, is derived. Despite this auspicious beginning, the remainder of Lamouroux's system was not based on pigmentation and as a result neither this nor that of the elder Agardh distinguished clearly between red, green and brown algae. Some years elapsed before Harvey (1836) divided the algae into four major divisions, solely on the basis of their pigmentation. This represents the first use of a biochemical criterion in plant systematics, a fact worth remembering when workers in other groups using 'modern' procedures make disparaging remarks about algal systematics. Harvey's four divisions comprised the Rhodospermae (red algae), Melanospermae (brown algae), Chlorospermae (green algae) and Diatomaceae. In general, Harvey's assignment of genera was reasonably accurate by modern standards although the considerable colour variation in many red algae produced some problems. The most serious of these was that *Porphyra* and *Bangia* were placed in the green algae rather than the red because of the frequent greenish colour of thalli of these two genera. The relationship between these two genera (and their relatives) and the other red algae was not appreciated for many years. By comparison, the relationship between the coralline algae and the red algae was established at a relatively early date. Although most pre-Linnaean workers had regarded the coralline algae as plants, Linnaeus (1758) was strongly influenced by the work of Ellis (1755, 1767) and followed the latter's views that they should be considered as animals. This opinion was continued for many years and, as late as 1816, Lamouroux still regarded the coralline algae together with other calcified algae, as animal corals. The change from the Linnaean interpretation came first for those coralline algae where the thallus is jointed and articulate which were treated as plants by Schweigger (1819) and Gray (1821). The encrusting genera now assigned to the Corallinaceae were not accepted as plants until the somewhat later work of Phillipi (1837).

The use of colour as a systematic criterion has been extended throughout the algae so that pigmentation is currently a fundamental criterion in algal systematics and several other major categories, in addition to those proposed by Harvey, are accepted on this basis at the present time.

The systematics of the red algae as a whole

The incorrect placement of *Bangia* and *Porphyra* in the green algae by Harvey (see p. 232) delayed association of these algae and their allies with the remaining red algae for many years. The studies of Berthold (1882b) were responsible for the alignment of these two groups of algae even though many of the observations which led to this association have not been confirmed and some must be regarded with extreme suspicion (see p. 164). The differences between *Bangia*, *Porphyra* and their allies on the one hand and the remaining red algae on the other are such that the two groups were considered by the beginning of the present century as subclasses of the class Rhodophyceae and named the Bangioideae and Florideae respectively. The name Protoflorideae was suggested for the former (Rosenvinge, 1909) and has appeared in various texts although, in view of its phylogenetic connotations, it is an unfortunate choice; however, being a later synonym of Bangioideae it can be rejected on grounds of priority. The names used for the two subclasses were not in accordance with the recommendations of the nomenclatural code and were changed to Bangiophycidae and Florideophycidae respectively (Newton, 1953). One trend during the present century has been to raise the status of all major groups of algae and to treat these as Divisions (= phyla) rather than simply as classes, so that the class Rhodophyceae was more recently regarded as a division, Rhodophyta, and the two major subdivisions were considered as classes rather than as subclasses (Melchior, 1954; Copeland, 1956; Cronquist, 1960). The terms Bangiophyceae and Florideophyceae

234 BIOLOGY OF THE RHODOPHYTA

are due to Melchior (1954) and Cronquist (1960) respectively. There are four possible arrangements involving major and minor categories, as follows:

(1) **Class** Rhodophyceae
   Subclass Bangioideae
   Subclass Florideae

(3) **Class** Rhodophyceae
   Subclass Protoflorideae
   Subclass Florideae

(3) **Class** Rhodophyceae
   Subclass Bangiophycideae
   Subclass Florideophycideae

(4) **Division** Rhodophyta
   Class Bangiophyceae
   Class Florideophyceae

Of these, the first two must be dismissed on purely nomenclatural grounds and there are left the third and fourth alternatives. The fundamental question is whether the major category should be regarded as a class or as a division and, in consequence, whether the two minor categories should be treated as subclasses or as classes. The similarities between the two minor categories are considerable in terms of the principal criteria used in the major classification of the algae. The chemical and physiological similarities of the phycobilin pigments has been one of the most significant criteria for the establishment of the group 'red algae' while the carotenoid pigments are virtually identical in the two groups. The structure of the cell walls, with unorientated cellulose microfibrils and cellulose molecules with low crystallinity, is similar in both groups while the mucilages are chemically similar. The carbohydrate reserve materials are virtually identical, with floridean starch, galactan sulphate polymers, floridoside and trehalose present in both although mannoglyceric acid has never been reported in the Bangiophyceae. The structural similarities are also considerable at the fine structural level, particularly in terms of the complete absence of flagella although the plug-like pit connections, once thought to be restricted to the major group, have been demonstrated also in relatives of *Bangia* (see p. 8). At higher morphological levels there are many differences, such as the essentially apical cell division of the major group and intercalary cell division in the minor. There are certain apparent similarities in certain aspects of reproduction on which the alignment of the two groups was based originally, of very dubious validity. These arguments suggest that the fourth arrangement is the most feasible for the red algae, based on the information available at the present time.

The systematics of the two classes, Bangiophyceae and Florideophyceae, currently accepted in the Rhodophyta, developed almost independently of one another and it would seem best to treat the two classes separately in any further discussion, taking first the Florideophyceae.

Florideophyceae

The subsequent development of the systematics of the Florideophyceae depended on increased knowledge of thallus morphology, reproductive structures and life histories. Studies of Florideophycean anatomy and morphology prior to the acceptance of the cell concept make curious reading today and the wonder is that authors such as Dillwyn (1802-1809) were able to prepare such careful, critical and detailed analyses of the algae under consideration in complete ignorance of the cell. Once this fundamental concept was established, numerous morphological studies, by Nägeli and the younger Agardh, etc., provided a much more solid basis by which to compare organisms than the previous accounts based almost entirely on external features. Knowledge of reproductive structures developed much more slowly. Red algal spermatangia and tetrasporangia were figured by the middle of the eighteenth century (Ellis, 1767) but it was a century before the gamete fusion was demonstrated (Bornet and Thuret, 1867) in a number of Florideophyceae and a further 40 years before the sequence of nuclear phases was established (Yamanouchi, 1906) and the role of the tetrasporangium elucidated. The crucial step in the establishment of the present systematics came with detailed developmental studies of the carposporophyte. Although the 'cystocarp' of the Florideophyceae had been defined by the younger Agardh as early as 1844 and then used extensively by him in his diagnoses of new genera and species and expanded descriptions of long-known entities, he rarely used little more than the external form of the mature structure. It could well be that his failure to consider developmental stages was a consequence of J. Agardh's dependence on dried herbarium material and his very rare personal collection of fresh material in the field after 1845. Following the correct interpretation of gamete fusion in Florideophyceae by Bornet and Thuret (1867) various workers had examined post-fertilisation developments in a range of genera but it remained for Schmitz to clarify the resulting confusion. Schmitz (1883) showed that in a few Florideophyceae the carposporophyte develops directly from the carpogonium whereas in the majority the carpogonium forms cells or filaments which fuse with another cell and that it is from the latter that the carposporophyte develops. Schmitz termed the structure which participated in the second fusion the 'auxiliary cell' and although his interpretation that this represented a form of double fertilisation was disproved within a few years, the ontogenetic patterns which he demonstrated served as the basis for the new systematics.

The present classification of the Florideophyceae is derived from that proposed originally by Schmitz (1889, in Schmitz and Hauptfleisch, 1896-1897), in which four orders were recognised, Gigartinales, Rhodymeniales, Cryptonemiales and Nemalionales. It has been indicated (Christensen, 1967) that the last-named should be more correctly named Nemaliales and this orthography will be used in the remainder of the discussion whatever the original spelling. Schmitz based his system of classification on the structure and development of the carposporophyte, with particular emphasis on the position of the auxiliary cell. Considering the inadequate information available to Schmitz at the time his system was proposed, it is surprising that there should have been so few changes. Various revisions of the ordinal arrangement have been suggested during the past 70 years but of these only one is of major importance. The separation by Oltmanns (1904), as the order Ceramiales, of those members of the Rhodymeniales of Schmitz in which the auxiliary cell is formed *after* fertilisation has been shown to be well justified. Most of the remaining proposals are not of any great significance. Proposals by Kylin (1925) creating the order Nemastomales (more correctly, Nemastomatales) for the families Nemastomaceae (Nemastomataceae) and Rhodophyllidaceae, placed previously in the orders Cryptonemiales and Gigartinales respectively, and by Sjöstedt (1926) of the order Sphaerococcales for the family Sphaerococcaceae, placed previously in the Gigartinales, were not accepted by subsequent authors and the two new orders were reduced by Kylin (1928a), the constituent families being placed in the Gigartinales. One area which has been the subject of much controversy concerns the reclassification of the order Nemaliales. The Gelidiaceae was the first family to be separated and given ordinal status as the Gelidiales (Kylin, 1923). Subsequently, it has been suggested that the Bonnemaisoniaceae, placed originally by Schmitz in the Rhodymeniales but transferred by Kylin (1916a) to the Nemaliales should be given ordinal status (Feldmann and Feldmann, 1942). It has also been suggested that the Acrochaetiaceae be given ordinal status (Chemin, 1937; Feldmann, 1953) while Desikachary (1958, 1963) elevated the Chaetangiaceae to ordinal status as the Chaetangiales. Thus, at the present time, the most conservative system for the Florideophyceae (Kylin, 1956) accepts six orders, Nemaliales, Gelidiales, Cryptonemiales, Gigartinales, Rhodymeniales and Ceramiales while there are these various recent proposals regarding the Nemaliales and involving either its further subdivision or the re-amalgamation with it of the Gelidiaceae and the elimination of the order Gelidiales.

Of the six orders, the Ceramiales is probably the most clearly circumscribed. The uniformity of thallus development in each of the four families, Ceramiaceae, Delesseriaceae, Rhodomelaceae and Dasyaceae, and particularly the stereotyped development of the carposporophyte in them are remarkable. There is some evidence of variability in the formation of the auxiliary cell (Dixon, 1964) but no other indications of deviation from a standard pattern of development. Opinion of the status of the Rhodymeniales is somewhat divided. This order has been described as 'one of the most sharply defined orders' (Papenfuss, 1966) while Drew (1954a) has commented that in her opinion there is no fundamental difference between certain families of the Gigartinales, such as the Rhodophyllidaceae, Hypneaceae and Gracilariaceae, and those of the Rhodymeniales. The indefinite nature of the distinction between the Cryptonemiales and certain families of the Gigartinales has been appreciated for many years (Fritsch, 1945) while it has been stated (Papenfuss, 1951) that 'the Cryptonemiales and Gigartinales are especially marked by seemingly overlapping types, but whether the types represent a true relationship or merely instances of parallel evolution cannot at present be determined'.

The situation with respect to the Nemaliales and the relationship between this order and the Gelidiales must take into account the mass of new information which has been accumulated regarding life histories and reproductive development in the genera involved. Taking first the arguments for the separation of the Gelidiaceae from the Nemaliales and its elevation to ordinal status, this was based initially (Kylin, 1923) on the supposed 'diplobiontic' life history, the evidence for which was derived entirely from the occurrence of tetrasporangial as well as gametangial phases. Subsequently (Kylin, 1928a), it was concluded that an auxiliary cell, in the Kylin sense of the term, did not occur so that *Gelidium* resembled members of the Nemaliales and differed from the other orders of the Florideophyceae. Considering life histories in the Nemaliales there is now abundant evidence to indicate that the supposed 'haplobiontic' life history attributed to members of this order has no justification. Regarding the development of the carposporophyte in members of the Gelidiaceae, studies (Dixon, 1959; Fan, 1961) of various genera and species have disclosed that the carposporophyte does not develop directly from the carpogonium as described by Kylin (1928a) but from an inflated cell which is formed from the carpogonium. There is, however, some divergence of opinion as to how this inflated cell develops. According to Fan, fusion with the hypogynous cell is obligatory and fusions with other cells can then occur, whereas Dixon states that the inflated cell can develop in certain instances without any fusion at all, or that the carpogonium may fuse with the hypogynous cell, or with neighbouring cells, or with both hypogynous cell and neighbouring cells. The differences between these two accounts are not very great although they have considerable significance with respect to the subsequent interpretation in that acceptance of an auxiliary cell in these genera of the Gelidiaceae, as elsewhere, depends on whether the fusions are or are not obligatory. On the basis of Fan's data, the hypogynous cell is an auxiliary cell in *Gelidium* whereas this is not the case with the information presented by Dixon. When Kylin first separated the Gelidiaceae from the Nemaliales, the latter was apparently very homogeneous. Subsequent investigations have disclosed that the various families of the Nemaliales, such as the Batrachospermaceae, Helminthocladiaceae, Chaetangiaceae, Acrochaetiaceae, Naccariaceae and Bonnemaisoniaceae are all well defined, each family forming a discrete group of genera, but with little to hold them together as an order, unless one accepts the very restricted definition of an auxiliary cell proposed by Kylin for this very purpose. Initially Kylin had argued that the Nemaliales was characterised by the absence of an auxiliary cell but when investigations of *Bonnemaisonia* and *Asparagopsis* disclosed such a structure, Kylin added a further requirement that the auxiliary cell could not be a cell of the carpogonial branch, despite the acceptance of an auxiliary cell in this position elsewhere in the Florideophyceae. As mentioned previously (p. 237) elevation of various families such as the Acrochaetiaceae,

Bonnemaisoniaceae, Chaetangiaceae to ordinal status has been proposed. There have been innumerable publications dealing with these proposals, which have been summarised by Dixon (1961) and Papenfuss (1966), with almost every possible combination of family and order. There is adequate evidence of the heterogeneity of the families grouped in the Nemaliales but no adequate basis at the present time either for the separation of these as independent orders or for the separation of the Gelidiaceae. The only difference between the views of Dixon and Papenfuss is that the latter does not agree with the return of the Gelidiaceae to the Nemaliales, an opinion to which the present author must take exception.

Two very different proposals which diverge considerably from the classical Schmitz through Kylin historical development are those of Stoloff and Chadefaud. Stoloff (1962), dissatisfied with the use of mere morphological criteria, attempted to use the biochemical criteria associated with the polysaccharide content of various red algae. Although the type of polysaccharide present was correlated to some extent with the ordinal status adopted on the basis of the Kylin scheme, this was not always the case and a new order Hypneales was created to accommodate those entities which diverged. The procedures for chemical analysis of polysaccharides are now much more advanced than ten years ago and the Stoloff proposals must be re-analysed critically on the basis of these improved procedures before it is possible to make any final pronouncements as to the virtue of polysaccharide chemistry as a systematic criterion. This approach is discussed in more detail with respect to the future developments likely in the systematics of the Rhodophyta.

The Chadefaud scheme involved the division of the Florideophyceae into three groups which were termed 'Eo-floridées', 'Meso-floridées' and 'Meta-floridées', within which nine orders were grouped, as follows (1960):

Eo-floridées	Acrochaetiales Eu-Nemalionales
Meso-floridées	Chaetangiales Gelidiales Gigartinales Cryptonemiales Rhodymeniales
Meta-floridées	Bonnemaisoniales Ceramiales

Although, at first sight, the Chadefaud scheme represents a major switch in emphasis it consists of little more than a rearrangement of the families of the Nemaliales. Of the nine orders listed, five (Gelidiales, Gigartinales, Cryptonemiales, Rhodymeniales and Ceramiales) do not differ in any respect to the same entities as treated by Kylin (1956), while the remaining four (Acrochaetiales, Eu-Nemalionales, Chaetangiales, and Bonnemaisoniales) are subdivisions of Nemaliales. The nomenclature adopted by Chadefaud for the three supra-ordinal categories is not in accord with the Code of Botanical Nomenclature while the names themselves are *nomina nuda*, being devoid of diagnoses.

In the present treatment of the Florideophyceae, only five orders are accepted, Nemaliales, Gigartinales, Cryptonemiales, Rhodymeniales and Ceramiales. The assignment of families and genera to these, insofar as those entities to which reference is made in the present text, is listed in a subsequent section.

Bangiophyceae

Although certain parts of the systematic arrangement of the Florideophyceae as generally accepted at the present time are still a matter for controversy, as has been indicated in the preceding section, the system is founded on a considerable body of data, much of which has been confirmed and repeated many times. By comparison, the systematics of the Bangiophyceae are in an extremely primitive state because of the minimal amount of information which is available for many of the entities involved. As is indicated elsewhere (pp. 45, 164), many unsolved problems exist not only with respect to the most elementary details of morphological structure and thallus development but also relative to such fundamental aspects of reproduction and life history as the nature of a spore and the existence of sexual processes.

The initial incorrect placement of *Porphyra* and *Bangia* with the green algae persisted, despite suggestions (Thuret, 1855; Rabenhorst, 1868) that these algae were more closely allied to the red algae, until the more detailed investigations of Berthold (1882b) provided evidence, however doubtful some of this might now appear (see p. 174), of closer affinity with the Rhodophyceae, as they were then termed. Thus, at the beginning of the present century the algae now referred to the Bangiophyceae were placed in a single order, the Bangiales, in the subclass of the Rhodophyceae to which the nomenclaturally incorrect name Bangioideae was applied. Little further development occurred until the study by Kylin (1937b) which separated the genera with unicellular thalli as the order Porphyridiales, with the remaining multi-cellular members of the group in the order Bangiales. This sub-division was followed shortly after by the major revision by Skuja (1939) in which the group was divided further into four orders, Por-phyridiales, Goniotrichales, Bangiales and Compsopogonales with a fifth order, the Rhodochaetales, strongly supported if not formally proposed. The major criteria for the segregation of these orders was the basic morphology of the thallus so that the Skuja system may be characterised as follows:

Porphyridiales: thalli unicellular, occurring either as unicells or as irregular colonial masses; no evidence for sexual reproduction.

Goniotrichales: thalli filamentous, branched, or unbranched, fila-ments arising by intercalary cell division; no evidence for sexual reproduction.

Bangiales: thalli multicellular, filamentous, tubular or plate-like, with growth through intercalary cell division.

Compsopogonales: thalli filamentous or tubular, branched.

Rhodochaetales: thallus filamentous, uniseriate, composed of elongate cells formed by the transverse division of an apical cell.

This scheme has been accepted by all subsequent authors (Papenfuss, 1955; Kylin, 1956) with little modification during the past 30 years.

It has been shown (p. 48) that it is extremely difficult to distinguish between some of the unicellular members, referred by Skuja to the Porphyridiales, when these occur in colonial masses, and some of the 'filamentous' members assigned by him to the Goniotrichales where the cell aggregation is somewhat weak. The relationship between specimens referred to *Chroothece* (Porphyridiales) and *Asterocytis* (Goniotrichales) is probably the best example known where two entities referred to the two orders appear to merge morphologically. Recent work (Lewin and Robertson, 1971) indicating that a marine *Asterocytis* changed its morphology from filamentous to a unicellular or bicellular form resembling *Chroothece* with reduction in salinity of the growth medium from full to one quarter seawater is further evidence of the unsatisfactory nature of this discriminatory criterion. Feldmann (1955) proposed that the two orders should be merged because of this *Chroothece/Asterocytis* relationship and this appears to be justified. This merger would assign three families to the Porphyridiales (*sensu lato*), the Porphyridiaceae, Goniotrichaceae and the Phragmonemataceae. The only other change from the original Skuja scheme is the addition of the new family Boldiaceae to the Bangiales, in order to accommodate the newly described genus *Boldia* (Herndon, 1964).

In the present treatment of the Bangiophyceae, only four orders are accepted, Porphyridiales, Bangiales, Compsopogonales and Rhodochaetales. The assignment of families and genera to these, insofar as those entities to which reference is made in the present text, is considered in a subsequent section.

Future developments in red algal systematic studies

The declared goals of modern taxonomy have been outlined (Davis and Heywood, 1963) as follows:

1. To provide a convenient method of identification and communication.
2. To provide a classification which as far as possible expresses the natural relationships of organisms.
3. To detect evolution at work, discovering its processes and interpreting its results.

Knowledge of the world's flora is extremely uneven although increased understanding is occurring, sometimes rapidly, sometimes extremely slowly. It has been indicated (Davis and Heywood, 1963) that the development of taxonomic knowledge occurs in four phases. The first is the pioneer, or exploratory, phase which is concerned primarily with identification, with data derived from restricted material, often available only in a dried state. The second phase is the consolidation phase in which increased knowledge leads to the elimination of many of the arbitrary judgments of the first phase and results in the reduction of many names to synonymy. The third, or biosystematic, phase follows acquisition of cytological or biosystematic information and the emphasis changes more towards microevolution. The fourth phase represents an encyclopaedic co-ordination of the preceding three phases. In the Angiosperms, classification has moved some distance from the pioneer phase for many organisms or groups in particular geographical areas such as Europe or North America although the red algae are still very much at the pioneer phase. Knowledge of red algae is extremely scanty in many parts of the world, particularly for the Indian and Pacific Oceans and, even for European or North American taxa that are often thought to be well known, the lack of information is often surprising. It is not generally appreciated how little is actually known about most marine or freshwater algae and the Rhodophyta is probably the worst known of any algal group.

The major reason for this lack of taxonomic understanding is that a pronounced aversion to taxonomic studies arose in botany at the beginning of the present century. This was a particularly critical time for phycology in that the pre-Darwinian concept that any morphological variant was a distinct taxon was only just becoming challenged for the Rhodophyta. Some phycologists, such as W. H. Harvey, had a profound comprehension of the variation in appearance of algae growing under different conditions or at different seasons. Unfortunately, such an outlook was rare due to most work being undertaken by continental European phycologists with little or no personal knowledge of the living plant and largely ignorant of the extreme variability of the algal thallus because of their lack of field experience. This situation was made worse by the general lack of critical studies in the algae so that casual remarks became accepted as a consequence of their frequent repetition. One feature of this uncritical attitude is that the application of names in phycology has been largely a matter of tradition even within such a restricted geographical area as the British Isles.

The species is generally accepted as the basic unit of taxonomy and although it might be questioned whether the species concept is the most effective method for the delimitation of working units, the alternatives are even more inferior at the present time. Definition of the term 'species' has rightly achieved a degree of notoriety in recent years although it does not appear to have been appreciated that the different ways in which the term has been defined have been largely determined by the aim of a particular investigator. The primary functions of classification, systematics and taxonomy are strictly utilitarian, as was indicated in the declared goals listed above. The simplest approach at the pioneer stage is to consider the species merely as a means for providing a convenient method for identification and communication. Such an approach has obvious limitations. In order to provide some degree of regularity for the delimitation of the working units, information must be available on the ways by which such units have come into being if a rational basis is to be provided for their circumscription. Speciation may be defined as genetic differentiation producing a minimum degree of irreversible evolutionary change or divergence that has constant recognisable morphological differences. This definition indicates clearly that the study of speciation requires information not only at the morphological but also at the cytological and biosystematic

I

levels. These levels are interconnected by the series: gene-enzyme-process, the interaction of various processes giving rise to a pattern that is manifested in the final form of the plant. The major difficulty preventing investigation of speciation in Rhodophyta is that most of the necessary data are simply not available. Furthermore, it is clear from the comments made previously on the cytology of the Rhodophyta (see p. 21) that the essential information can never be obtained without radical improvements in technique.

If then, as has been indicated, the taxonomy of most algae is still at the pioneer phase, what is needed for further advancement? One obvious requirement is an increase in knowledge of algae from areas hitherto unknown or underinvestigated. The ignorance that prevails at the present time with regard to floristic knowledge in parts of Africa, Asia, and South America is a reflection of the limited number of indigenous investigators in those areas, their often inadequate training, and the lack of resources available to them. The elimination of deficiencies such as these should be a major aim of international phycology at the present time. Although there is considerable interest in short-term 'raids' into territories which are underdeveloped phycologically these are really of little importance except in the earliest stages of the pioneer phase. A second major requirement at the species level, particularly, is to consider in great detail the problems of phenotypic variation. The sedentary habit of most algae increases susceptibility to changing environmental factors. The external form of the red algal thallus is still a major criterion for taxonomic discrimination despite the variability which is at last becoming accepted by phycologists. However, before one can move away from pioneer phase taxonomy, a considerable amount of information is needed on seasonal and environmental changes in the form of the thallus. Such investigations demand long-term sequential investigations and it is principally for this reason that short-term visits to areas which are phycologically underdeveloped are unsuitable. In terms of the causes for variation in the form of the thallus, the general principles underlying morphogenesis in the Rhodophyta have been discussed in detail previously (see p. 93). The detailed considerations of the taxonomic and systematic implications of data such as these are needed for the second 'consolidation' phase of taxonomic development in which the arbitrary judgments of the primary pioneer phase are gradually replaced by more logical approaches in which the consequences of morphological plasticity are fully appreciated.

Considering next those categories higher than the species, improvements in higher systematics will only be achieved by a more phylogenetic approach. This will never be achieved unless features other than the purely morphological are taken into consideration. Despite the fact that the first use in the plant kingdom of a biochemical criterion for systematic discrimination was in connection with the Rhodophyta, most phylogenetic discussions have restricted themselves to considerations only of morphology. Various attempts are now being made to enlarge the factual basis of red algal systematics particularly by the use of biochemical criteria. Essentially, the major thrusts are in the direction either of carbohydrate chemistry or of protein chemistry with minor studies related to the occurrence of specific compounds.

Of the biochemical criteria based on protein chemistry, this approach was essentially that involved in the context of immunological studies first used for red algae by Steinecke (1925) and incorporated into a phylogenetic scheme by Wilke (1929). This work attempted to show, albeit at a very crude level but still somewhat revolutionary for the time, the extent of similarity or difference between representative members of the various orders. Because the procedure was so novel, phycologists were either unwilling or unable to appreciate the significance of this approach. As a consequence, the conclusions were either dismissed as being of no consequence (Fritsch, 1945) or ignored completely (Kylin, 1956) in discussions of red algal phylogeny. A more exact immunological approach based on a consideration of specific proteins was developed by Vaughan (unpublished observations). This attempted to relate the protein moieties of the phycobilin pigments of red algae. The results were of considerable value but somewhat limited by the difficulties involved in maintaining animals and sera. Difficulties such as these and the time-consuming aspects of immunological studies will always reduce the potential of this method.

Spectrophotometric analysis of the phycobilin pigments has been the approach used in various investigations (Hirose and Kumano, 1966; Hirose et al., 1969) which have attempted to assess the variations within the Rhodophyta and the relations between the pigments of these algae and those of the Cyanophyta. Although there are indications of differences between the absorption spectra of the various orders of red algae, the results are still far from satisfactory if detailed phylogenetic and systematic relationships are to be detected. A more exact procedure is to analyse the soluble protein components of the cell by means of polyacrylamide gel electrophoresis (Mallery and Richardson, 1971).

I*

The results of such a procedure are most spectacular, particularly if the protein moieties of the phycobilin pigments are compared. It is possible to compare the banding patterns and it was shown that the least complex patterns were found with genera of the Bangiophyceae and the most complex patterns with genera of the Ceramiales, with patterns of an intermediate nature in members of the Nemaliales. Although analyses of this sort are available only for a very restricted number of genera, the degree of precision is of the necessary order of magnitude to provide extremely useful data for both phylogenetic and systematic considerations.

The various different carbohydrate polymers present in the red algae have been known for many years (see p. 2) although it is only recently that precise methods of analysis have become available, such as the bacterial procedure for selective breakdown (Yaphe, 1959). Various investigators (Stoloff and Silva, 1957; Stoloff, 1962) have attempted to make comparisons between the various major taxonomic groupings and the three major types of carbohydrate polymers of the red algae. Although certain patterns are indicated by these studies, the most recent investigations of carbohydrate chemistry using even more precise methods of analysis have indicated that the basic statements of the various supposed categories of carbohydrate need to be revised. Agar, for instance, is not simply a mixture of agarose and agaropectin as was formerly thought, but rather a complex mixture of polysaccharides all having the same basic backbone structure although substituted to a variable degree with charged groups (Duckworth and Yaphe, 1971a, b). Thus, although the initial conclusions relating taxonomy and systematics with carbohydrate chemistry appeared promising these conclusions need to be re-evaluated in the light of this more recent information.

Of the minor studies which involve the occurrence of specific compounds in species, genera or family groups, there are several which are promising but none which are sufficiently well established for their precise use in systematic discrimination. Floridorubin, for example, has been identified in only four genera of the Florideophyceae but these are all referred to the *Amansia* group of the Rhodomelaceae (Saenger *et al.*, 1969). Brominated phenols occur widely in the same family but also with a restricted distribution and these materials might also be usable as a biochemical criterion of systematic importance (Augier, 1953, 1967). Other compounds, such as mannoglyceric acid, occur widely but in such quantities that the only systematic relationship is in the proportion in which a particular material occurs. This is of much less use in systematics than some substance which is either present or absent. It is possible that the detailed biochemical studies of various Japanese workers (Ito *et al.*, 1966a, 1966b, 1967) by which various particular amino acids have been isolated and identified will be of use in the future but a considerable amount of further work is required before the full significance of these observations can be evaluated.

In conclusion, it can be stated that in general, although the need for biochemical criteria in systematic studies is understood, there is as yet insufficient information for this approach to be of final value in discrimination. Until such criteria are available, phylogenetic specu-lation based entirely on morphological criteria should be kept to a minimum.

Phylogenetic relationships of the Rhodophyta

Most speculation on the evolutionary origin and relationships of the red algae and of the various classes and groups within that division can be dismissed as mere flights of phylogenetic fancy. The basic reason for rejecting much of the information which has been presented is that it represents attempts to deduce the major branches of the phylogenetic tree simply from a consideration of its ultimate branchlets. In the absence of a detailed fossil record, studies based merely on the comparative morphology of recent forms are unlikely to be of much assistance. Suggestions have been made elsewhere (p. 245) as to the way in which such phylogenetic and systematic studies might be pursued, by the use of detailed biochemical characteristics and com-parative studies of protein chemistry involving amino acid sequences. At the present time, the data available is totally inadequate to enable the detailed relationships within the division to be indicated and no attempt will be made here to discuss the various suggestions which have been put forward (Papenfuss, 1951, 1955; Kylin, 1956).

With respect to the relationships of the red algae to the other algal divisions, a recent study (Klein and Cronquist, 1967) considers the data available and evaluates it critically. Despite the mass of informa-tion which has been obtained, the situation with regard to the origin and relationships of the red algae as a whole is still decidedly vague. The general conclusion is that the evidence available suggests that the red algae were probably derived from the blue-green algal line although it is admitted that many phycologists disagree with such an interpreta-tion. If one considers the factual evidence there are several chemical and structural features in common. Members of the blue-green and of both groups of red algae contain a full complement of respiratory cytochromes, chlorophyll a is present in all, while phycobilin pigments occur both in the red and blue-green algae. Phycocyanins predominate in the blue-green with phycoerythrins in the red algae. It is likely though that the same tetrapyrrolic phycobilin moieties occur in both although the proteins with which they are associated may be different. Of the carotenes, α-carotene is present in greater quantity than β-carotene in both red and blue-green algae, although there are several differences in the xanthophyll constituents. The carbohydrates of the blue-green and red algae are similar in a number of respects. Trehalose is common to the two divisions, while floridean starch and cyanophycean starch are not greatly dissimilar in that both are branched molecules, but with the latter less branched than the former. The cell wall chemistry of the red and blue-green algae shows certain points of similarity and several of difference. Cellulose microfibrils occur in both groups, although the degree of orientation of the fibrils is low compared with the green algae, for example, and xylans are more prominent than in the latter. The cell wall slimes in both red and blue-green algae are pentose sugars rarely found in other groups of algae. It may be concluded therefore that the biochemical similarities between red and blue-green algae are far greater in number than the points of difference. Of the structural features, the total absence of flagella in red and blue-green algae, although negative evidence, is of significance, in that flagella based on a $9+2$ formula are present in all other groups of plants. The presence of nuclei bounded by a membrane and their absence in the blue-green algae is one of the strongest arguments against a relationship between these two groups. The reports of 'amitosis' in red algae are probably nothing more than a consequence of the persistent nuclear envelope. One of the most significant questions requiring investigation concerns the presence or absence of centrioles in the red algae. The early studies of red algal cytology reported the occurrence of 'polar bodies' or 'evanescent centrosomes' in members of the Bangiophyceae and Florideophyceae but not consistently. West-brook (1935) observed polar bodies in only two of the genera investigated and then only rarely while Magne (1964a) commented on the variation in appearance. Knowledge of the fine structure is limited: the one published account (McDonald, 1972) comments that the object described here as the polar body resembles a centriole in shape and diameter but is shorter in length and lacks the 'cartwheel' substructure characteristic of the centrioles of other plants.   The relationship between the photosynthetic organelles in red and blue-green algae is similar to that between the nuclei, in that in red algae these are aggregated into organised chloroplasts with a double bounding membrane whereas in blue-green algae they lie free in the cytoplasm.   In addition, pyrenoids are present in most of the Bangiophyceae and some of the Florideophyceae whereas they are completely unreported in the blue-green algae.

The relationship between the red algae and other eucaryotic algae is still extremely doubtful and there are two alternative hypotheses.   It may be concluded that the membrane bound nuclei and chloroplasts evolved independently in the red and green algae, in which case there is no close relationship between these two groups, or it may be surmised that the eucaryotic state evolved only once from a blue-green ancestor and that there was then an early divergence to give the red algae on one side and the green algae and land plant derivatives on the other. The available evidence at the present time is insufficient to decide between these two alternatives, although a detailed investigation of the centriole or polar body in red algae might provide the critical evidence for the decision.

Early investigators commented many times on a possible relationship between the red algae and the ascomycete fungi.   The original argument was based largely upon a supposed resemblance between the female gametes in these two groups (Dodge, 1914) although a recent study of the pit connections (Lee, 1971a) has again suggested certain similarities between these two groups.   Although there are certain superficial features in common, the statement by Fritsch (1945) that 'such resemblances as have been emphasised appear to me to be more of the nature of parallel development and not to afford evidence of any phylogenetic connection' seems a more than adequate verdict.

Summary of orders, families and genera of the Rhodophyta

The following outline includes only those orders, families and genera to which reference is made in the present text.   The sequence and assignment are based on those of Kylin (1956) and deviations from this work are indicated.   Space does not permit detailed descriptions of the various categories in the present list and those interested in such information are referred to Kylin's synoptic treatment.

RHODOPHYTA

BANGIOPHYCEAE

PORPHYRIDIALES (including Goniotrichales of Kylin)
Porphyridiaceae: *Chroothece, Petrovanella, Porphyridium, Rhodella, Rhodospora, Vanhoffenia.*
Goniotrichaceae (separated with Phragmonemataceae as an order Goniotrichales by Kylin): *Asterocytis, Goniotrichum, Goniotrichopsis, Neevea.*
Phragmonemataceae (separated with Goniotrichaceae in the order Goniotrichales by Kylin): *Cyanoderma, Kyliniella, Phragmonema.*

BANGIALES
Erythropeltidaceae: *Erythrocladia, Erythropeltis, Erythrotrichia, Porphyropsis, Smithora.*
Bangiaceae: *Bangia, Porphyra, Porphyrella.*
Boldiaceae (Herndon, 1964): *Boldia.*

COMPSOPOGONALES
Compsopogonaceae: *Compsopogon.*

RHODOCHAETALES
Rhodochaetaceae: *Rhodochaete.*

FLORIDEOPHYCEAE

NEMALIALES (as Nemalionales in Kylin, and including Gelidiales of Kylin)
Acrochaetiaceae (as Chantransiaceae in Kylin): *Acrochaetium, Audouinella, Rhodochorton, Yamadaella.*
Lemaneaceae: *Lemanea.*
Batrachospermaceae: *Batrachospermum, Sirodotia.*
Helminthocladiaceae: *Cumagloia, Dermonema, Helminthocladia, Helminthora, Liagora, Nemalion, Trichogloea.*
Chaetangiaceae: *Galaxaura, Pseudogloiophloea, Scinaia.*

Gymnocodiaceae (an entirely fossil group, omitted by Kylin): *Permocalculus, Gymnocodium.*

Naccariaceae: *Atractophora, Naccaria.*

Bonnemaisoniaceae: *Asparagopsis, Bonnemaisonia, Delisia, Ptilonia.*

Gelidiaceae (separated as an order, Gelidiales, by Kylin): *Gelidium, Pterocladia, Suhria.*

CRYPTONEMIALES

Dumontiaceae: *Acrosymphyton, Cryptosiphonia, Dudresnaya, Dumontia, Gibsmithea, Pikea, Thuretellopsis.*

Weeksiaceae (included in Dumontiaceae by Kylin): *Weeksia.*

Gloiosiphoniaceae: *Gloiosiphonia, Schimmelmannia.*

Endocladiaceae: *Endocladia.*

Rhizophyllidaceae: *Desmia, Ochtodes, Rhizophyllis.*

Peyssoneliaceae (as Squamariaceae in Kylin): *Contarinea, Erythrodermis, Peyssonelia.*

Hildenbrandiaceae: *Hildenbrandia.*

Corallinaceae: *Bossiella, Calliarthron, Choreonema, Corallina, Jania, Lithoporella, Lithothamnium, Lithothrix, Melobesia, Phymatolithon, Porolithon, Sporolithon.*

Solenoporaceae (an entirely fossil family, omitted by Kylin) *Solenopora.*

Cryptonemiaceae (as Grateloupiaceae in Kylin): *Cryptonemia, Grateloupia, Halymenia, Lobocolax, Pachymeniopsis, Prionitis.*

Kallymeniaceae (as Callymeniaceae in Kylin): *Erythrophyllum, Kallymenia.*

Choreocolacaceae (as Choreolaceae in Kylin): *Choreocolax, Harveyella, Holmsella.*

GIGARTINALES

Cruoriaceae: *Cruoria, Cruoriopsis, Petrocelis.*

Calosiphoniaceae: *Bertholdia.*

Nemastomataceae (as Nematomaceae in Kylin): *Nemastoma, Platoma, Schizymenia.*

Sebdeniaceae: *Sebdenia.*

Gracilariaceae: *Gracilaria.*

Plocamiaceae: *Plocamium.*

Furcellariaceae: *Furcellaria, Halarachnion, Neurocaulon.*

Solieriaceae: *Agardhiella, Gardneriella, Opuntiella, Turnerella.*

Rabdoniaceae: *Catenella.*

Rhodophyllidaceae: *Calliblepharis, Cystoclonium, Rhodophyllis.*
Polyideaceae (placed in Cryptonemiales by Kylin): *Polyides.*
Phyllophoraceae: *Ahnfeltia, Gymnogongrus, Phyllophora, Stenogramme.*
Gigartinaceae: *Besa, Chondrus, Gigartina, Iridaea.*

RHODYMENIALES

Rhodymeniaceae: *Botryocladia, Chrysymenia, Coelarthrum, Coelothrix, Fauchea, Halosaccion, Rhodymenia.*
Champiaceae (as Lomentariaceae in Kylin): *Champia, Chylocladia, Coeloseira, Gastroclonium, Lomentaria.*

CERAMIALES

Ceramiaceae: *Antithamnion, Aristothamnion, Bornetia, Callithamnion, Ceramium, Compsothamnion, Corynospora, Crouania, Dohrniella, Griffithsia, Pleonosporium, Plumaria, Ptilota, Ptilothamnionopsis, Seirospora, Spermothamnion, Tiffaniella.*
Rhodomelaceae: *Bostrychia, Chondria, Colacopsis, Laurencia, Lenormandia, Polysiphonia, Pterosiphonia, Rhodomela, Rytiphlaea, Vidalia.*
Delesseriaceae: *Cryptopleura, Delesseria, Hemineura, Hypoglossum, Martensia, Nitophyllum, Phycodrys, Polyneura.*
Dasyaceae: *Dasya, Heterosiphonia, Rhodoptilum.*

BIBLIOGRAPHY

ABBOTT, I. A. 1946. The genus *Griffithsia* (Rhodophyceae) in Hawaii. *Farlowia*, **2**, 439-453.

ABBOTT, I. A. 1967. Studies in the foliose red algae of the Pacific coast. II. *Schizymenia*. *Bull. Sth. Calif. Acad. Sci.* **66**, 161-174.

ABBOTT, I. A. 1970. *Yamadaella*, a new genus in the Nemaliales (Rhodophyta). *Phycologia*, **9**, 115-123.

ADEY, W. H. & JOHANSEN, H. W. 1972. Morphology and taxonomy of Corallinaceae with special reference to *Clathromorphum, Mesophyllum*, and *Neopolyporolithon gen. nov.* (Rhodophyceae, Cryptonemiales). *Phycologia*, **11**, 159-180.

ADEY, W. H. & MCKIBBIN, D. L. 1970. Studies on the maerl species *Phymatolithon calcareum* (Pallas) *nov. comb.* and *Lithothamnium coralloides* Crouan in the Ria de Vigo. *Botanica mar.* **13**, 100-106.

AGARDH, C. A. 1817. *Synopsis algarum Scandinaviae.* . . . Berlingianus, Lund.

AGARDH, C. A. 1820-1828. *Species algarum* . . ., vols 1 (1), 1 (2), 2 (1). Mauritius, Lund.

AIRTH, R. L. & BLINKS, L. R. 1956. A new phycoerythrin from *Porphyra naiadum*. *Biol. Bull. mar. biol. Lab.*, *Woods Hole*, **111**, 321-327.

ASANO, M., KURODA, M. & ASAHI, T. 1951. Seasonal variations of the chemical composition of *Gelidium amansii* Lmx., collected from Usu. *Bull. Hokkaido reg. Fish. Res. Lab.* **1**, 28-35.

AUGIER, H. 1965a. Les substances de croissance chez la Rhodophycée *Botryocladia botryoides* (Wulf.) J. Feldm. *C. r. hebd. Séanc. Acad. Sci., Paris*, **260**, 2304-2306.

AUGIER, H. 1965b. Contribution à l'étude des facteurs de croissance des algues rouges. *Bull. Inst. oceanogr. Monaco*, **65** (1341), 1-18.

AUGIER, H. 1971. Contribution à l'étude de la répartition des phytohormones dans le thalle de la Rhodophycée *Botryocladia botryoides* (Wulf.) J. Feldm. *Tethys*, **3**, 183-188.

AUGIER, J. 1953. La constitution chimique de quelques Floridées Rhodomélacées. *Revue gén. Bot.* **60**, 257-283.

AUGIER, J. 1967. Biochimie et taxinomie chez les algues. *Mém. Soc. bot. Fr.* 1965, 8-15.

AUSTIN, A. P. 1956. Chromosome counts in the Rhodophyceae. *Nature, Lond.* **178**, 370-371.

AUSTIN, A. P. 1960a. Observations on the growth, fruiting and longevity of *Furcellaria fastigiata* (L.) Lam. *Hydrobiologia*, **15**, 193-207.

AUSTIN, A. P. 1960b. Life history and reproduction of *Furcellaria fastigiata* (L.) Lam. 2. The tetrasporophyte and reduction division in the tetrasporangium. *Ann. Bot.*, N.S. **24**, 296-310.

AUSTIN, A. P. & PRINGLE, J. D. 1968. Mitotic index in selected red algae *in situ*. I. Preliminary survey. *J. mar. biol. Ass. U.K.* **48**, 609-635.

AUSTIN, A. P. & PRINGLE, J. D. 1969. Periodicity of mitosis in red algae. *Proc. Int. Seaweed Symp.* **6**, 41-52.

BAARDSETH, E. 1941. The marine algae of Tristan da Cunha. *Res. Norw. sci. Exped. Tristan da Cunha 1937-8*, **9**, 1-174.

BARBER, H. N. 1947. Genetics and algal life cycles. *Aust. J. Sci.* **9**, 217-218.

BARTON, E. S. 1891. On the occurrence of galls in *Rhodymenia palmata* Grev. *J. Bot. Lond.* **29**, 65-68.

BATTEN, L. 1923. The genus *Polysiphonia* Grev., a critical revision of the British species, based upon anatomy. *J. Linn. Soc. (Bot.)*, **46**, 271-311.

BAUCH, R. 1937. Die Entwicklung der Bisporen der Corallinaceen. *Planta*, **26**, 365-390.

BELCHER, J. H. 1960. Culture studies of *Bangia atropurpurea* (Roth) Ag. *New Phytol.* **59**, 367-373.

BELCHER, J. H. & SWALE, E. M. F. 1960. Some British freshwater material of *Asterocytis*. *Br. phycol. Bull.* **2**, 33-35.

BERTHOLD, G. 1881. Zur Kenntniss der Siphoneen und Bangiaceen. *Mitt. zool. Stn Neapel*, **2**, 72-82.

BERTHOLD, G. 1882a. Beiträge zur Morphologie und Physiologie der Meeresalgen. *Jb. wiss. Bot.* **13**, 569-717.

BERTHOLD, G. 1882b. Die Bangiaceen des Golfes von Neapel und der angrenzenden Meeresabschnitte. *Fauna Flora Golf. Neapel*, **8**, 1-28.

BIEBL, R. 1937. Oekologische und zellphysiologische Studien an Rotalgen der englischen Südküste. *Beih. bot. Zbl.* A **57**, 381-424.

BIEBL, R. 1962. Seaweeds. In: *Physiology and biochemistry of algae*, ed. by Lewin, R. A., 799-815. Academic Press, New York & London.

BISALPUTRA, T. & BISALPUTRA, A. A. 1967. The occurrence of DNA fibrils in chloroplasts of *Laurencia spectabilis*. *J. Ultrastruct. Res.* **17**, 14-22.

BLINKS, L. R. 1954. The photosynthetic function of pigments other than chlorophyll. *A. Rev. Pl. Physiol.* **5**, 93-114.

BLINKS, L. R. 1955. Photosynthesis and productivity of littoral marine algae. *J. mar. Res.* **14**, 363-373.

BLINKS, L. R.   1965.   Accessory pigments and photosynthesis.   In: *Photophysiology*, vol. 1, ed. by Giese, A. C., 199-221.   Academic Press, New York & London.

BOCQUET, C.   1953.   Sur un Copépode harpacticoide mineur, *Diarthrodes feldmanni* n. sp.   *Bull. Soc. Zool. Fr.* **78**, 101-105.

BODARD, M.   1968.   L'infrastructure des 'corps en cerise' des *Laurencia* (Rhodomelacées, Ceramiales).   *C. r. hebd. Séanc. Acad. Sci., Paris*, sér. D **266**, 2393-2396.

BODARD, M.   1972.   Étude morphologique et cytologique d'*Helminthocladia senegalensis* (Rhodophycées), Nemalionale nouvelle à carpotetraspores et à cycle haplodiplophasique.   *Phycologia*, **10**, 361-374.

BOILLOT, A.   1958.   Sur la présence en France d'un *Compsopogon* (Rhodophyceae, Bangioideae).   *Bull. Soc. phycol. Fr.* **4**, 13-15.

BOILLOT, A.   1965.   Sur l'alternance de générations hétéromorphes d'une Rhodophycée, *Halarachnion ligulatum* (Woodward) Kützing (Gigartinales, Furcellariacées).   *C. r. hebd. Séanc. Acad. Sci., Paris*, **261**, 4191-4193.

BONEY, A. D.   1960.   Observations on the spore output of some common red algae.   *Br. phycol. Bull.* **2**, 36-37.

BONEY, A. D.   1965.   Aspects of the biology of seaweeds of economic importance.   *Adv. mar. Biol.* **3**, 105-235.

BONEY, A. D.   1966.   *A biology of marine algae*.   Hutchinson, London.

BONEY, A. D.   1967.   Spore emission, sporangium proliferation and spore germination *in situ* in the monosporangia of *Acrochaetium virgatulum*.   *Br. phycol. Bull.* **3**, 317-326.

BØRGESEN, F.   1927.   Marine algae of the Canary Islands. III. Rhodophyceae.   Part I.   Bangiales and Nemalionales.   *Biol. Meddr*, **6** (6), 1-97.

BORNET, E.   1892.   Les algues de P.-K.-A. Schousboe.   *Mém. Soc. natn. Sci. nat. math. Cherbourg*, **28**, 165-376.

BORNET, E. & THURET, G.   1867.   Recherches sur la fécondation des Floridées.   *Annls Sci. nat. (Bot.)*, ser. 5, **7**, 137-166.

BORNET, E. & THURET, G.   1876-1880.   *Notes algologiques*, vols 1, 2. Masson, Paris.

BOUCK, G. B.   1962.   Chromatophore development, pits, and other fine structure in the red alga *Lomentaria baileyana* (Harv.) Farlow.   *J. Cell Biol.* **12**, 553-569.

BOURRELLY, P.   1970.   *Les algues d'eau douce*, vol. 3.   Boubee, Paris.

BOWER, F. O.   1908.   *The origin of a land flora*.   Macmillan, London.

BOWER, F. O.   1922.   The primitive spindle as a fundamental feature in the embryology of plants.   *Proc. R. Soc. Edinb.* **43**, 1-36.

BRADY, G. S.   1894.   On *Fucitrogus rhodymeniae*, a gall-producing Copepod.   *Jl R. microsc. Soc.* 1894, 168-170.

BRODY, M. & BRODY, S. S.   1962.   Light reactions in photosynthesis.

K

In: *Physiology and Biochemistry of Algae*, ed. by Lewin, R. A., 3-23. Academic Press, New York & London.

BRONGNIART, A. 1829. *Histoire des végétaux fossiles*, vol. 1. Dufour & d'Ocagne, Paris.

BRUCKER, W. 1958. Zur Bildung von Tumoren an Meeresalgen. I. *Arch. Geschwulstforsch.* **14**, 10-12.

BUGNON, F. 1961. Sur l'existence de thalles parenchymateux chez les Floridées. *Bull. Soc. bot. Fr.* **108**, 2-31.

BÜNNING, E. 1935. Zellphysiologische Studien an Meeresalgen. *Protoplasma*, **22**, 444-456.

BUTLER, M. R. 1936. Seasonal variations in *Chondrus crispus*. *Biochem. J.* **30**, 1338-1340.

CABIOCH, J. 1969. Les fonds de maerl de la Baie de Morlaix et leur peuplement végétale. *Cah. Biol. mar.* **10**, 139-161.

CABIOCH, J. 1972. Étude sur les Corallinacées. *Cah. Biol. mar.* **12**, 121-286.

CAMPBELL, D. H. 1940. *The evolution of land plants*. Stanford University Press, Stanford.

CHADEFAUD, M. 1952. Sur le cycle sexuel des organisms eucaryotes et son évolution. *Revue scient., Paris*, **90**, 49-57.

CHADEFAUD, M. 1960. *Traité de Botanique systematique*, vol. 1. Masson, Paris.

CHADEFAUD, M. 1962. Sur quelques détails de l'organisation morphologique des parois cellulaires chez les Floridées filamenteuses. *Bull. Soc. bot. Fr.* **109**, 148-156.

CHAMBERS, J. E. 1966. *Some electron microscope observations on Rhodophycean fine structure*. Ph.D. thesis, University of Kansas.

CHAPMAN, D. J. & CHAPMAN, V. J. 1961. Life histories in the algae. *Ann. Bot.*, N.S. **25**, 547-561.

CHAPMAN, V. J. 1970. *Seaweeds and their uses*, 2nd edition. Methuen, London.

CHAVE, K. E. 1954. Aspects of the biogeochemistry of magnesium. I. Calcareous marine organisms. *J. Geol.* **62**, 266-283.

CHEMIN, E. 1928. Multiplication végétative et dissémination chez quelques algues Floridées. *Trav. Stn biol. Roscoff*, **7**, 5-62.

CHEMIN, E. 1937. Le développement des spores chez les Rhodophycées. *Revue gén. Bot.* **49**, 205-234; 300-327; 353-374; 424-448; 478-536.

CHESTER, G. D. 1896. Notes concerning the development of *Nemalion multifidum*. *Bot. Gaz.* **21**, 340-347.

CHIHARA, M. 1960. On the germination of tetraspores of *Falkenbergia hillebrandii* (Bornet) Falkenberg. *J. Jap. Bot.* **35**, 249-253.

CHIHARA, M. 1962. Life cycle of the Bonnemaisoniaceous algae in Japan (2). *Scient. Rep. Tokyo Kyoiku Daigaku*, sect. B **11**, 27-54.

CHIHARA, M.   1965.   Germination of the carpospores of *Bonnemaisonia nootkana*, with special reference to the life cycle.   *Phycologia*, **5**, 71-79.

CHILD, C. M.   1916.   Axial susceptibility gradients in algae.   *Bot. Gaz.* **62**, 89-114.

CHRISTENSEN, T.   1967.   Two new families and some new names and combinations in the algae.   *Blumea*, **15**, 91-94.

CHURCH, A. H.   1919.   Historical review of the Florideae.   *J. Bot., Lond.* **57**, 297-304; 329-334.

CLAUS, G.   1961.   Beiträge zur Kenntnis der Algenflora der ababligeten Höhlen.   *Hydrobiologia*, **19**, 192-222.

CODOMIER, L.   1969.   Sur l'alternance de générations hétéromorphes du *Neurocaulon grandifolium* Rodriguez (Rhodophycée, Gigartinale).   *C. r. hebd. Séanc. Acad. Sci., Paris*, sér. D **269**, 1060-1062.

CONRAD, H. & SALTMAN, P.   1962.   Growth substances.   In: *Physiology and biochemistry of algae*, ed. by Lewin, R. A., 663-671.   Academic Press, New York & London.

CONWAY, E.   1964.   Autecological studies of the genus *Porphyra*. 1. The species found in Britain.   *Br. phycol. Bull.* **2**, 342-348.

COOPER, N. C. & JOHNSTONE, G. R.   1944.   The seasonal production of agar in *Gelidium cartilagineum*, a perennial red alga.   *Am. J. Bot.* **31**, 638-640.

COPELAND, H. F.   1956.   *The classification of lower organisms*.   Pacific Books, Palo Alto.

COTTON, A. D.   1912.   Marine algae.   *Proc. R. Ir. Acad.* **31** (15), 1-178.

COUTÉ, A.   1971.   Sur le cycle morphologique du *Liagora tetrasporifera* comparé à celui du *Liagora distenta* (Rhodophycées, Némalionales, Helminthocladiacées).   *C. r. hebd. Séanc. Acad. Sci., Paris*, sér. D **273**, 626-629.

CRONQUIST, A.   1960.   The divisions and classes of plants.   *Bot. Rev.* **26**, 425-482.

CRONSHAW, J., MYERS, A. & PRESTON, R. D.   1958.   A chemical and physical investigation of the cell walls of some algae.   *Biochim. biophys. Acta*, **27**, 89-103.

CROUAN, P. L. & CROUAN, H. M.   1867.   *Florule du Finistère*.   Klincksieck, Paris.

DANFORTH, W. F.   1962.   Substrate assimilation and heterotrophy.   In: *Physiology and biochemistry of algae*, ed. by Lewin, R. A., 99-123. Academic Press, New York & London.

DANGEARD, P.   1927.   Recherches sur les *Bangia* et les *Porphyra*.   *Botaniste*, **18**, 183-244.

DANGEARD, P.   1931.   Sur le développement des spores chez quelques *Porphyra*.   *Trav. Crypt. dedies L. Mangin*, 85-96.

DANGEARD, P.   1940.   Recherches sur les enclaves iridescentes de la cellule des algues. *Botaniste*, **31**, 31-63.

DANGEARD, P. 1949. Les algues marines de la côte occidentale du Maroc. *Botaniste*, **34**, 89-189.

DANGEARD, P. 1954. Contribution à la connaisance du cycle évolutif des Bangiacées. *Proc. int. Bot. Congr.* **8** (17), 76-79.

DARBISHIRE, O. V. 1898. *Bangia pumila* Aresch., eine endemische Alge der östlichen Östee. *Wiss. Meeresunters., Abt. Kiel*, N.F. **3**, 27-31.

DARLINGTON, C. D. 1937. *Recent advances in cytology*. Churchill, London.

DAS, C. R. 1963. The Compsopogonales in India. *Proc. natn. Inst. Sci. India*, **29B**, 239-243.

DAVIS, P. H. & HEYWOOD, V. H. 1963. *Principles of angiosperm taxonomy*. Oliver & Boyd, Edinburgh & London.

DAWES, C. J., SCOTT, F. M. & BOWLER, E. 1961. A light- and electron-microscopic survey of algal cell walls. Phaeophyta and Rhodophyta. *Am. J. Bot.* **48**, 925-934.

DEBAUX, O. 1873. Algues marines du Corse. *Recl Mém. Méd. Chir. Pharm. milit.* 1873, 528-541.

DENIZOT, M. 1968. *Les algues Floridées encroutantes (à l'exclusion des Corallinacées)*. Museum National d'Histoire Naturelle, Paris.

DERBÉS, A. & SOLIER, A. J. J. 1856. Mémoire sur quelques points de la physiologie des algues. *C. r. hebd. Séanc. Acad. Sci., Paris* Suppl. **1**, 1-120.

DESIKACHARY, T. V. 1958. Taxonomy of algae. *Mem. Indian bot. Soc.* **1**, 52-62.

DESIKACHARY, T. V. 1962. *Cumagloia* Setchell *et* Gardner and *Dermonema* (Grev.) Harv. *J. Indian bot. Soc.* **41**, 132-147.

DESIKACHARY, T. V. 1963. Status of the order Chaetangiales (Rhodophyta). *J. Indian bot. Soc.* **42A**, 16-26.

DE VALÉRA, M. & FOLAN, A. 1964. Germination *in situ* of carpospores in Irish material of *Asparagopsis armata* Harv. and *Bonnemaisonia asparagoides* (Woodw.) Ag. *Br. phycol. Bull.* **2**, 332-338.

DILLWYN, L. W. 1802-1809. *British Confervae.* . . . W. Phillips, London.

DIXON, P. S. 1959. The structure and development of the reproductive organs and carposporophyte in two British species of *Gelidium*. *Ann. Bot.*, N.S. **23**, 397-407.

DIXON, P. S. 1960. Studies on marine algae of the British Isles. The genus *Ceramium*. *J. mar. biol. Ass. U.K.* **39**, 315-374.

DIXON, P. S. 1961. On the classification of the Florideae with particular reference to the position of the Gelidiaceae. *Botanica mar.* **3**, 1-16.

DIXON, P. S. 1963a. The taxonomic implications of the 'pit connexions' reported in the Bangiophyceae. *Taxon*, **12**, 108-110.

259

DIXON, P. S. 1963b. Variation and speciation in marine Rhodophyta. In: *Speciation in the sea*, ed. by Harding, J. P. & Tebble, N., 51-62. Systematics Association, London.

DIXON, P. S. 1963c. Terminology and algal life histories, with particular reference to the Rhodophyta. *Ann. Bot.*, N.S. **27**, 353-355.

DIXON, P. S. 1963d. The Rhodophyta: some aspects of their biology. *Oceanogr. mar. Biol. ann. Rev.* **1**, 177-196.

DIXON, P. S. 1964. Auxiliary cells in the Ceramiales. *Nature, Lond.* **201**, 519-520.

DIXON, P. S. 1965. Perennation, vegetative propagation and algal life histories, with special reference to *Asparagopsis* and other Rhodophyta. *Botanica gothoburg.* **3**, 67-74.

DIXON, P. S. 1966a. On the form of the thallus in the Florideophyceae. In: *Trends in plant morphogenesis*, ed. by E. Cutter, 45-63. Longmans, Green, London.

DIXON, P. S. 1966b. The Rhodophyceae. In: *The chromosomes of the algae*, ed. by Godward, M. B. E., 168-204. Edward Arnold, London.

DIXON, P. S. 1970. The Rhodophyta: some aspects of their biology. II. *Oceanogr. mar. Biol. ann. Rev.* **8**, 307-352.

DIXON, P. S. 1971a. A study of *Callithamnion lejolisea* Farl. *J. Phycol.* **7**, 58-63.

DIXON, P. S. 1971b. Cell enlargement in relation to the development of thallus form in Florideophyceae. *Br. phycol. J.* **6**, 195-205.

DIXON, P. S., MURRAY, S. N., RICHARDSON, W. N. & SCOTT, J. L. 1972. Life histories in the Cryptonemiales. *Mém. Soc. bot. Fr.* (in press).

DIXON, P. S. & RICHARDSON, W. N. 1969a. The life history of *Thuretellopsis peggiana* Kylin. *Br. phycol. J.* **4**, 87-89.

DIXON, P. S. & RICHARDSON, W. N. 1969b. The life histories of *Bangia* and *Porphyra* and the photoperiodic control of spore production. *Proc. Int. Seaweed Symp.* **6**, 133-139.

DIXON, P. S. & RICHARDSON, W. N. 1970. Growth and reproduction in red algae in relation to light and dark cycles. *Ann. N.Y. Acad. Sci.* **175**, 764-777.

DODGE, B. O. 1914. The morphological relationships of the Florideae and the Ascomycetes. *Bull. Torrey bot. Cl.* **41**, 157-202.

DOTY, M. S. 1971. Physical factors in the production of tropical benthic marine algae. In: *Fertility of the sea*, vol. 1, ed. by Costlow, J. D., 99-121. Gordon & Breach, New York.

DOUBT, D. G. 1935. Notes on two species of *Gymnogongrus*. *Am. J. Bot.* **22**, 294-310.

DREW, K. M. 1934. Contributions to the cytology of *Spermothamnion turneri* (Mert.) Aresch. I. The diploid generation. *Ann. Bot.*, N.S. **48**, 549-573.

DREW, K. M. 1937. *Spermothamnion snyderae* Farlow, a Floridean alga bearing polysporangia. *Ann. Bot.*, N.S. **1**, 463-476.

DREW, K. M. 1939. An investigation of *Plumaria elegans* (Bonnem.) Schmitz with special reference to triploid plants bearing parasporangia. *Ann. Bot.*, N.S. **3**, 347-367.

DREW, K. M. 1943. Contributions to the cytology of *Spermothamnion turneri* (Mert.) Aresch. II. The haploid and triploid generations. *Ann. Bot.*, N.S.

DREW, K. M. 1944. Nuclear and somatic phases in the Florideae. *Biol. Rev.* **19**, 105-120.

DREW, K. M. 1945. A hitherto undescribed form of adult axis in the genus *Batrachospermum*. *Nature, Lond.* **155**, 608.

DREW, K. M. 1952. Studies in the Bangioideae. I. Observations on *Bangia fuscopurpurea* (Dillw.) Lyngb. in culture. *Phytomorphology*, **2**, 38-51.

DREW, K. M. 1954a. The organization and inter-relationships of the carposporophytes of living Florideae. *Phytomorphology*, **4**, 55-69.

DREW, K. M. 1954b. Studies in the Bangioideae. III. The life-history of *Porphyra umbilicalis* (L.) Kütz. var. *laciniata* (Lightf.) J. Ag. *Ann. Bot.*, N.S. **18**, 183-211.

DREW, K. M. 1955. Life histories in the algae with special reference to the Chlorophyta, Phaeophyta and Rhodophyta. *Biol. Rev.* **30**, 343-390.

DREW, K. M. 1956. Reproduction in the Bangiophycidae. *Bot. Rev.* **22**, 553-611.

DRING, M. J. 1967. Effects of daylength on growth and reproduction on the *Conchocelis* of *Porphyra tenera*. *J. mar. biol. Ass. U.K.* **47**, 501-510.

DUCKWORTH, M. & YAPHE, W. 1971a. The structure of agar. Part I. Fractionation of a complex mixture of polysaccharides. *Carbohyd. Res.* **16**, 189-197.

DUCKWORTH, M. & YAPHE, W. 1971b. The structure of agar. Part II. The use of a bacterial agarase to elucidate structural features of the charged polysaccharides in agar. *Carbohyd. Res.* **16**, 435-445.

DUYSENS, L. N. M. 1951. Transfer of light energy within the pigment systems present in photosynthesizing cells. *Nature, Lond.* **168**, 548-550.

EDELSTEIN, T. 1970. The life history of *Gloiosiphonia capillaris* (Hudson) Carmichael. *Phycologia*, **9**, 55-59.

ELLIOT, G. F. 1955. The Permian calcareous alga *Gymnocodium*. *Micropalaentology*, **1**, 83-90.

ELLIS, J. 1755. *An essay towards a natural history of the Corallines.* Ellis, London.

ELLIS, J. 1767. On the animal nature of the genus of Zoophytes called *Corallina*. *Phil. Trans. R. Soc.* **57**, 404-427.

ENGELMANN, T. W. 1883. Farbe und Assimilation. *Bot. Ztg*, **41**, 1-13.

ERNST, J. 1958. The life-forms of some perennial marine algae of Roscoff and their vertical distribution. *Abs. Int. Seaweed Symp.* **3**, 31.

FAN, K. C. 1961. Morphological studies of the Gelidiales. *Univ. Calif. Publs Bot.* **32**, 315-368.

FELDMANN, G. 1965. Le développement des tétraspores de *Falkenbergia rufolanosa* et le cycle des Bonnemaisoniales. *Revue gén. Bot.* **72**, 621-626.

FELDMANN, G. 1970a. Sur l'ultrastructure des corps irisants des *Chondria* (Rhodophycées). *C. r. hebd. Séanc. Acad. Sci.*, *Paris*, sér. D **270**, 945-946.

FELDMANN, G. 1970b. Sur l'ultrastructure de l'appareil irisant du *Gastroclonium clavatum* (Roth) Ardissone (Rhodophycées). *C. r. hebd. Séanc. Acad. Sci.*, *Paris*, sér. D **270**, 1244-1246.

FELDMANN, J. 1937. Recherches sur la végétation marine de la Méditerranée. La Côte des Alberes. *Revue algol.* **10**, 1-339.

FELDMANN, J. 1951. Ecology of marine algae. In: *Manual of phycology*, ed. by Smith, G. M., 313-334. Chronica Botanica, Waltham.

FELDMANN, J. 1952a. Développement du carposporophyte chez le *Bertholdia neapolitana* (Berthold) Schmitz. *C. r. hebd. Séanc. Acad. Sci.*, *Paris*, **234**, 2552-2554.

FELDMANN, J. 1952b. Les cycles de reproduction des algues et leurs rapports avec la phylogénie. *Revue Cytol. Cytophysiol. vég.* **13**, 1-49.

FELDMANN, J. 1953. L'évolution des organes femelles chez les Floridées. *Proc. Int. Seaweed Symp.* **1**, 11-12.

FELDMANN, J. 1955. Un nouveau genre de Protofloridée: *Colacodictyon*, nov. gen. *Bull. Soc. bot. Fr.* **102**, 23-28.

FELDMANN, J. 1966. Les types biologiques d'algues marines benthiques. *Mém. Soc. bot. Fr.* 1966, 45-60.

FELDMANN, J. & FELDMANN, G. 1939a. Sur le développement des carpospores et l'alternance de générations de l'*Asparagopsis armata* Harvey. *C. r. hebd. Séanc. Acad. Sci.*, *Paris*, **208**, 1240-1242.

FELDMANN, J. & FELDMANN, G. 1939b. Sur l'alternance de générations chez les Bonnemaisoniacées. *C. r. hebd. Séanc. Acad. Sci.*, *Paris*, **208**, 1425-1427.

FELDMANN, J. & FELDMANN, G. 1942. Recherches sur les Bonnemaisoniacées et leur alternance de générations. *Annls Sci. nat. (Bot.)*, sér. **11**, 3, 75-175.

FELDMANN, J. & FELDMANN, G. 1946. A propos d'un récent travail du Prof. Kylin sur l'alternance de générations du *Bonnemaisonia asparagoides*. *Bull. Soc. Hist. nat. Afr. N.* **37**, 35-38.

FELDMANN, J. & FELDMANN, G. 1948. Sur l'existence de synapses secondaires chez une Céramiacée. *Bull. Soc. Hist. nat. Afr. N.* **39**, 125-128.

FELDMANN, J. & FELDMANN, G. 1950. Les 'corps en cerise' du *Laurencia obtusa* (Huds.) Lamour. *C. r. hebd. Séanc. Acad. Sci.*, Paris, **231**, 1335-1337.

FELDMANN, J. & FELDMANN, G. 1952. Nouvelles recherches sur le cycle des Bonnemaisoniacées: le développement des tétraspores du *Falkenbergia rufolanosa* (Harv.) Schmitz. *Revue gén. Bot.* **59**, 313-323.

FELDMANN, J. & FELDMANN, G. 1970. Sur l'ultrastructure des synapses des algues rouges. *C. r. hebd. Séanc. Acad. Sci.*, Paris, sér. D, **271**, 292-295.

FELDMANN, J. & MAZOYER, G. 1937. Sur l'identité de l'*Hymenoclonium serpens* (Crouan) Batters et du protonéma du *Bonnemaisonia asparagoides* (Woodw.) C. Ag. *C. r. hebd. Séanc. Acad. Sci.*, Paris, **205**, 1084-1085.

FELDMANN, J. & TIXIER, R. 1947. Sur la floridorubine, pigment rouge des plastes d'une Rhodophycée (*Rytiphlaea tinctoria* (Clem.) C. Ag.). *Revue gén. Bot.* **54**, 341-353.

FELDMANN-MAZOYER, G. 1941. *Recherches sur les Céramiacées de la Méditerranée occidentale.* Minerva, Algiers.

FLINT, L. H. 1947. Studies of freshwater red algae. *Am. J. Bot.* **34**, 125-131.

FLINT, L. H. 1953. *Kyliniella* in America. *Phytomorphology*, **3**, 76-80.

FLINT, L. H. 1955. *Hildenbrandia* in America. *Phytomorphology*, **5**, 185-189.

FRITSCH, F. E. 1942. Studies in the comparative morphology of the algae. I. Heterotrichy and juvenile stages. *Ann. Bot.*, N.S. **6**, 397-412.

FRITSCH, F. E. 1945. *Structure and reproduction of the algae*, vol. 2. Cambridge University Press, Cambridge.

FUJIYAMA, T. 1957. (Cytochemical study on the crown-gall of *Porphyra tenera*) (in Japanese). *Univ. Tokyo Publs Fish. Sci.* (*Amemiya Jubilee Volume*), 829-840.

FUNK, G. 1927. Die Algenvegetation des Golfs von Neapel. *Pubbl. Staz. zool. Napoli* **7** (suppl.), 1-507.

FUNK, G. 1955. Beiträge zur Kenntnis der Meeresalgen von Neapel. *Pubbl. Staz. zool. Napoli*, **25** (suppl.), 1-178.

GANTT, E. & CONTI, S. F. 1965. The ultrastructure of *Porphyridium cruentum*. *J. Cell Biol.* **26**, 365-381.

GANTT, E. & CONTI, S. F. 1966. Granules associated with the chloroplast lamellae of *Porphyridium cruentum*. *J. Cell Biol.* **29**, 423-434.

GANTT, E., EDWARDS, M. R. & CONTI, S. F. 1968. Ultrastructure of *Porphyridium aerugineum*, a blue-green colored Rhodophytan. *J. Phycol.* **4**, 65-71.

GENSE, M. T., GUÉRIN-DUMARTRAIT, E., LECLERC, J. C. & MIHARA, S. 1969. Synchronisation de *Porphyridium*. Évolution des quantités de pigments et de la capacité photosynthétique au cours du cycle biologique. *Phycologia*, **8**, 135-141.

GESSNER, F. 1970. Temperature—Plants. In: *Marine ecology*, vol. 1 (1), ed. by Kinne, O., 363-406. Wiley-Interscience, London.

GESSNER, F. & SCHRAMM, W. 1971. Salinity—Plants. In: *Marine ecology*, vol. 1 (2), ed. by Kinne, O., 705-820. Wiley-Interscience, London.

GIBBS, S. P. 1970. The comparative ultrastructure of the algal chloroplast. *Ann N.Y. Acad. Sci.* **175**, 454-473.

GIRAUD, A. & MAGNE, F. 1968. La place de la méiose dans le cycle de développement de *Porphyra umbilicalis*. *C. r. hebd. Séanc. Acad. Sci.*, Paris, sér. D, **267**, 586-588.

GISLEN, T. 1930. Epibioses of the Gullmar fjord. *Skr. svenska Vetensk-Akad.* 1930 (4), 1-380.

GOEBEL, K. 1878. Zur Kenntnis einiger Meeresalgen. *Bot. Ztg*, **36**, 177-184.

GOTELLI, I. B. & CLELAND, R. E. 1968. Differences in the occurrence and distribution of hydroxproline-proteins among the algae. *Am. J. Bot.* **55**, 907-914.

GRAY, S. F. 1821. *Natural arrangement of British plants*, 2 vols. Baldwin, Craddock & Joy, London.

GREEN, P. B. 1960. Wall structure and lateral formation in the alga *Bryopsis*. *Am. J. Bot.* **47**, 476-481.

GREEN, P. B. 1962. Cell expansion. In: *Physiology and biochemistry of the algae*, ed. by Lewin, R. A., 625-632. Academic Press, New York & London.

GREGORY, B. D. 1934. On the life-history of *Gymnogongrus griffithsiae* Mart. and *Ahnfeltia plicata* Fries. *J. Linn. Soc.* (*Bot.*), **49**, 531-551.

GRUBB, V. M. 1923a. Preliminary note on the reproduction of *Rhodymenia palmata* Ag. *Ann. Bot.* **37**, 151-152.

GRUBB, V. M. 1923b. The attachments of *Porphyra umbilicalis* (L.) J. Ag. *Ann. Bot.* **37**, 131-140.

GRUBB, V. M. 1925. The male organs of the Florideae. *J. Linn. Soc.* (*Bot.*), **47**, 177-255.

GUILLARD, R. R. L. 1962. Salt and osmotic balance. In: *Physiology and biochemistry of algae*, ed. by Lewin, R. A., 529-540. Academic Press, New York & London.

HAMPSON, M. A. 1957. The laverbread industry in South Wales and the laverweed. *Fishery Invest., Lond.*, ser. II, **21** (7), 1-8.

HANIC, L. A. & CRAIGIE, J. S. 1969. Studies on the algal cuticle. *J. Phycol.* **5**, 89-102.

HARDER, R. 1923. Über die Bedeutung von Lichtintensität und Wellenlänge für die Assimilation farbiger Algen. *Z. Bot.* **15**, 305-355.

HARDER, R. 1948. Einordnung von *Trailliella intricata* in den Generationswechsel bie Bonnemaisoniaceae. *Nachr. Akad. Wiss. Göttingen*, 1948, 24-27.

HARDING, J. P. 1954. The copepod *Thalestris rhodymeniae* (Brady) and its nauplius parasitic in the seaweed *Rhodymenia palmata* (L.) Grev. *Proc. zool. Soc. London,* **124,** 153-161.

L'HARDY-HALOS, M. T. 1969a. La formation des anastomoses chez *Pleonosporium borreri* (Smith) Naegeli ex Hauck et *Bornetia secundiflora* (J. Ag.) Thuret (Rhodophyceae-Ceramiaceae). *C. r. hebd. Séanc. Acad. Sci., Paris,* sér. D, **268,** 276-278.

L'HARDY-HALOS, M. T. 1969b. La morphogénèse chez les Ceramiaceae: organisation hiérarchique de la fronde. *Mém. Soc. bot. Fr.* **115,** 142-148.

L'HARDY-HALOS, M. T. 1971a. Recherches sur les Céramiacées (Rhodophycées, Céramiales) et sur quelques aspects de leur morphogénèse. *Bull. Soc. scient. Bretagne,* **46,** 99-112.

L'HARDY-HALOS, M. T. 1971b. Recherches sur les Céramiacées (Rhodophycées-Céramiales) et leur morphogénèse. III. Observations et recherches expérimentales sur la polarité cellulaire et pour la hiérarchisation des éléments de la fronde. *Revue gén. Bot.* **78,** 407-491.

HARVEY, W. H. 1836. Algae. In: *Flora Hibernica,* 2 vols, ed. by Mackay, J. T. Part 2, 157-256. Curry, Dublin.

HASE, E. 1962. Cell division. In: *Physiology and biochemistry of algae,* ed. by Lewin, R. A., 617-624. Academic Press, New York and London.

HASSINGER-HUIZINGA, H. 1952. Generationswechsel und Geschlechtsbestimmung bei *Callithamnion corymbosum* (Sm.) Lyngb. *Arch. Protistenk,* **98,** 91-124.

HAUPTFLEISCH, P. 1897. Die als fossile Algen (und Bakterien) beschreibenen Pflanzenreste oder Abdrucke. In: *Die naturlichen Pflanzenfamilien* . . ., vol. 1 (2), ed. by Engler, A. & Prantl, K., 545-569. Engelmann, Leipzig.

HAXO, F. 1960. Photosynthesis in algae containing special pigments. In: *Handbuch der Pflanzenphysiologie,* vol. 5 (2) ed. by Ruhland, W., 349-363. Springer, Berlin.

HEEREBOUT, G. R. 1968. Studies on the Erythropeltidaceae (Rhodophyceae-Bangiophycidae). *Blumea,* **16,** 139-157.

HELLEBUST, J. A. 1970. Light—Plants. In: *Marine ecology,* vol. 1 (1), ed. by Kinne, O., 125-158. Wiley-Interscience, London.

HERNDON, W. R. 1964. *Boldia*: a new Rhodophycean genus. *Am. J. Bot.* **51,** 575-581.

HEYN, A. N. J. 1931. Der Mechanismus der Zellstreckung. *Recl. Trav. bot. néerl.* **28,** 113-244.

HIROSE, H. & KUMANO, S. 1966. Spectroscopic studies on the phycoerythrins from Rhodophycean algae with special reference to their phylogenetic relationships. *Bot. Mag., Tokyo,* **79,** 105-113.

HIROSE, H., KUMANO, S. & MADONO, K. 1969. Spectroscopic studies on phycoerythrins from Cyanophycean and Rhodophycean algae with special reference to their phylogenetic relationships. *Bot. Mag., Tokyo,* **82,** 197-203.

HÖFLER, K. 1931. Hypotonietod und osmotische Resistenz einiger Rotalgen. *Öst. bot. Z.* **80,** 51-71.

HOFMEISTER, W. 1851. *Vergleichende Untersuchungen der . . . hoherer Kryptogamen.* Hofmeister, Leipzig.

HOLLENBERG, G. J. 1940. New marine algae from Southern California. I. *Am. J. Bot.* **27,** 868-877.

HOLLENBERG, G. J. 1968. An account of the species of *Polysiphonia* of the central and western tropical Pacific Ocean. I. Oligosiphonia. *Pacif. Sci.* **22,** 56-98.

HOLLENBERG, G. J. & ABBOTT, I. A. 1968. New species of marine algae from California. *Can. J. Bot.* **46,** 1235-1251.

HONSELL, E. 1963. Prime osservazioni ultrastutturali sulle cellule di *Bangia fuscopurpurea* (Dillw.) Lyngb. *Delpinoa,* N.S. **5,** 139-155.

HOOGENHOUT, H. 1963. Synchronization of cell division in *Porphyridium aerugineum* Geitler. *Naturwissenschaften,* **50,** 456-457.

HOWE, M. A. 1914. The marine algae of Peru. *Mem. Torrey bot. Club,* **15,** 1-185.

HUMM, H. J. 1951. The red algae of economic importance. In: *marine products of commerce,* 2nd edition, ed. by Tressler, D. K. & Lemon, J. M., 47-93. Reinhold, New York.

HURDELBRINK, L. & SCHWANTES, H. O. 1972. Sur le cycle de développement d'un *Batrachospermum. Mém. Soc. Bot. Fr.* (in press).

HUS, H. T. A. 1902. An account of the species of *Porphyra* found on the Pacific coast of North America. *Proc. Calif. Acad. Sci.,* ser. 3 (*Bot.*), **2,** 173-240.

HYGEN, G. 1945. Life-cycles and nuclear phases in Florideae. *Nytt Mag. Naturvid.* **85,** 89-98.

ILLMAN, W. I., MCLACHLAN, J. & EDELSTEIN, T. 1970. Marine algae of the Champlain Sea episode near Ottawa. *Can. J. Earth Sci.* **7,** 1583-1585.

INOH, S. 1947. (*Development of algae*) (in Japanese). Inoh, Tokyo.

ISHIKAWA, M. 1921. Cytological studies on *Porphyra tenera* Kjellm. *Bot. Mag., Tokyo,* **35,** 206-218.

ITO, K. & HASHIMOTO, Y. 1966. Isolation of a new amino acid, 'Gigartinine', from a red alga, *Gymnogongrus flabelliformis. Bull. Jap. Soc. sci. Fish.* **32,** 274-279.

ITO, K., MIYAZAWA, K. & HASHIMOTO, Y. 1966. Distribution of gongrine and gigartinine in marine algae. *Bull. Jap. Soc. scient. Fish.* **32,** 727-729.

ITO, K., MIYAZAWA, K. & HASHIMOTO, Y. 1967. Occurrence of γ-guanidinobutyric acid and concentration of gongrine and gigartinine in a red alga, *Gymnogongrus flabelliformis*. *Bull. Jap. Soc. scient. Fish.* **33**, 572-577.

IWASAKI, H. 1965. Studies on the physiology and ecology of *Porphyra tenera*. *J. Fac. Fish. Anim. Husb. Hiroshima Univ.* **6**, 133-211.

IYENGAR, M. O. P. & BALAKRISHNAN, M. S. 1949. Morphology and cytology of *Polysiphonia platycarpa* Boergs. *Proc. Indian Acad. Sci.*, sect. B, **29**, 105-108.

JANET, C. 1914. *L'alternance sporophytogamétophytique de générations chez les algues*. Ducourtieux & Gout, Limoges.

JENNINGS, R. C. & McCOMB, A. J. 1967. Gibberellins in the red alga *Hypnea musciformis* (Wulf.) Lamour. *Nature, Lond.* **215**, 872-873.

JOFFÉ, R. 1896. Observations sur la fécondation des Bangiacées. *Bull. Soc. bot. Fr.* **43**, 143-146.

JOHNSON, J. H. 1960. Paleozoic Solenoporaceae and related red algae. *Colo. Sch. Mines Q.* **55** (3), 1-77.

JOHNSON, J. H. 1961. *Limestone-building algae and algal limestones*. Colorado School of Mines, Golden.

JONES, W. E. 1956. Effect of spore coalescence on the early development of *Gracilaria verrucosa* (Hudson) Papenfuss. *Nature, Lond.* **178**, 426-427.

JUNGERS, V. 1933. Recherches sur les plasmodesmes chez les végétaux. II. Les synapses des algues rouges. *Cellule*, **42**, 7-28.

KATADA, M. 1963. Life forms of sea-weeds and succession of their vegetation. *Bull. Jap. Soc. scient. Fish.* **29**, 798-808.

KATAYAMA, T. & FUJIYAMA, T. 1957. Studies on the nucleic acid of algae with special reference to the desoxyribonucleic acid contents of the crown-gall tissues developed on *Porphyra tenera* Kjellman. *Bull. Jap. Soc. scient. Fish.* **23**, 249-254.

KILLIAN, C. 1914. Über die Entwicklung einiger Florideen. *Z. Bot.* **6**, 209-278.

KINOSHITA, S. & TERAMOTO, K. 1958. On the efficiency of gibberellin on the growth of *Porphyra*-frond. *Bull. Jap. Soc. Phycol.* **6**, 85-88.

KINZEL, H. 1956. Untersuchungen über Bau und Chemismus der Zellwände von *Antithamnion cruciatum* (Ag.) Naeg. *Protoplasma*, **46**, 445-474.

KINZEL, H. 1960. Über den Bau der Zellwände von *Bornetia secundiflora* (J. Ag.) Thur. *Botanica mar.* **1**, 74-85.

KIRBY, R. H. 1953. Seaweeds in commerce. *Colon. Pl. Anim. Prod.* **1**, 183-216; 284-293: **2**, 1-22.

KITO, H. 1966. Cytological studies of several species of *Porphyra*. I. Morphological and cytological observations on a species of *Porphyra* epiphytic on *Grateloupia filicina* var. *porracea* (Mert.) Howe. *Bull. Fac. Fish. Hokkaido Univ.* **16**, 206-208.

KITO, H. 1967. Cytological studies of several species of *Porphyra*.

II. Mitosis in carpospore germlings of *Porphyra yezoensis*. *Bull. Fac. Fish. Hokkaido Univ.* 18, 201-202.

KITO, H., YABU, H. & TOKIDA, J. 1967. The number of chromosomes in some species of *Porphyra*. *Bull. Fac. Fish. Hokkaido Univ.* 18, 59-60.

KJELLMAN, F. R. 1883. The algae of the Arctic Sea. *K. svenska Vetensk-Akad. Handl.* 20 (5), 1-350.

KLEIN, R. M. & CRONQUIST, A. 1967. A consideration of the evolutionary and taxonomic significance of some biochemical, micromorphological, and physiological characters in the thallophytes. *Q. Rev. Biol.* 42, 105-296.

KNAGGS, F. W. 1965. Spermatangia on the tetrasporophyte of *Rhodochorton floridulum* (Dillw.) Näg. *Nova Hedwigia*, 10, 269-272.

KNAGGS, F. W. 1966a. *Rhodochorton purpureum* (Lightf.) Rosenvinge. Observations on the relationship between morphology and environment I. *Nova Hedwigia*. 10, 499-513.

KNAGGS, F. W. 1966b. *Rhodochorton purpureum* (Lightf.) Rosenvinge. Observations on the relationship between morphology and environment II. *Nova Hedwigia*, 11, 337-349.

KNAGGS, F. W. 1967. *Rhodochorton purpureum* (Lightf.) Rosenvinge. Observations on the relationship between morphology and environment III. *Nova Hedwigia*, 12, 521-528.

KNAGGS, F. W. 1970. A review of Florideophycidean life histories and of the culture techniques employed in their investigation. *Nova Hedwigia*, 18, 293-330.

KNAGGS, F. W. & CONWAY, E. 1964. The life history of *Rhodochorton floridulum* (Dillw.) Näg. I. Spore germination and the form of the sporelings. *Br. phycol. Bull.* 2, 339-341.

KNIEP, H. 1928. *Die Sexualität der neideren Pflanzen.* Fischer, Jena.

KNIGHT, M. & PARKE, M. W. 1931. *Manx algae.* Liverpool University Press, Liverpool.

KNOX, E. 1926. Some steps in the development of *Porphyra naiadum*. *Publs. Puget Sound mar. biol. Stn*, 5, 125-135.

KONRAD-HAWKINS, E. 1964. Developmental studies on regenerates of *Callithamnion roseum* Harvey. *Protoplasma*, 58, 42-74.

KORNMANN, P. 1961. Zur Kenntnis der *Porphyra*-Arten von Helgoland. *Helgoland. Wiss. Meeresunters.* 8, 176-192.

KOSCHTSUG, C. 1872. Entwicklungsgeschichte von *Callithamnion daviesii* Lyngb. und *Porphyra laciniata* Ag. *Abh. neurruss. Naturf. Ges.* 1, 17-21.

KREGER, D. R. 1962. Cell walls. In: *Physiology and biochemistry of algae*, ed. by Lewin, R. A., 315-335. Academic Press, New York London.

KRISHNAMURTHY, V. 1959. Cytological investigations on *Porphyra umbilicalis* (L.) Kütz. var. *laciniata* (Lightf.) J. Ag. *Ann. Bot.*, N.S. 23, 147-176.

KRISHNAMURTHY, V. 1962. The morphology and taxonomy of the genus *Compsopogon* Montagne. *J. Linn. Soc. (Bot.)*, 58, 207-222.

KUCKUCK, P. 1912. Über *Platoma bairdii* (Farl.) Kuck. *Wiss. Meeresunters.*, *Abt. Helgoland*, N.F. 5, 187-208.

KUGRENS, P. 1970. Comparative ultrastructure of several parasitic red algae. *J. Phycol.* 6 (suppl.), 9.

KUGRENS, P. & WEST, J. A. 1972. Synaptonemal complexes in red algae. *J. Phycol.* 8, 187-191.

KUNIEDA, H. 1939. On the life history of *Porphyra tenera* Kjellman. *J. Coll. Agric. imp. Univ. Tokyo*, 14, 377-405.

KUNZENBACH, R. & BRUCKER, W. 1960. Zur Bildung von 'Tumoren' an Meeresalgen II. *Ber. dt. bot. Ges.* 73, 8-18.

KÜTZING, F. T. 1843. *Phycologia generalis.* Brockhaus, Leipzig.

KYLIN, H. 1912. Studien über die schwedischen Arten der Gattungen *Batrachospermum* Roth und *Sirodotia* nov. gen. *Nova Acta R. Soc. Scient. upsal.*, ser. 4 3 (3), 1-40.

KYLIN, H. 1915. Untersuchungen über die Biochemie der Meeresalgen. *Hoppe-Seyler's Z. physiol. Chem.* 94, 337-425.

KYLIN, H. 1916a. Die Entwicklungsgeschichte und die systematische Stellung von *Bonnemaisonia asparagoides* (Woodw.) Ag. *Z. Bot.* 8, 545-586.

KYLIN, H. 1916b. Über die Befruchtung und Reduktionsteilung bei *Nemalion multifidum*. *Ber. dt. bot. Ges.* 34, 257-271.

KYLIN, H. 1917a. Über die Keimung der Florideensporen. *Ark. Bot.* 14 (22), 1-25.

KYLIN, H. 1917b. Über die Entwicklungsgeschichte von *Batrachospermum moniliforme*. *Ber. dt. bot. Ges.* 35, 155-164.

KYLIN, H. 1923. Studien über die Entwicklungsgeschichte der Florideen. *K. svenska VetenskAkad. Handl.* 63 (11), 1-139.

KYLIN, H. 1924. Studien über die Delesseriaceen. *Acta Univ. lund.*, N.F., Avd. 2, 20 (6), 1-111.

KYLIN, H. 1925. The marine red algae in the vicinity of the Biological Station at Friday Harbor, Wash. *Acta Univ. lund.*, N.F., Avd. 2, (9), 1-87.

KYLIN, H. 1928a. Entwicklungsgeschichtliche Florideenstudien. *Acta Univ. lund.*, N.F., 24 (4), 1-127.

KYLIN, H. 1928b. Über *Falkenbergia hillebrandii* und ihre Beziehung zur Aspaltung von Iod. *Bot. Notiser*, 1928, 233-254.

KYLIN, H. 1930a. Über die Entwicklungsgeschichte der Florideen. *Acta Univ. lund.*, N.F., Avd. 2, 26 (6), 1-104.

KYLIN, H. 1930b. Über die Blasenzellen bei *Bonnemaisonia*, *Trailliella* und *Antithamnion*. *Z. Bot.* 23, 217-226.

KYLIN, H. 1937a. Anatomie der Rhodophyceen. In: *Handbuch der Pflanzenanatomie*, Abt. 2, 6 (2), ed. by Linsbauer, K. i-vii+1-347, Borntraeger, Berlin.

KYLIN, H. 1937b. Über eine marine *Porphyridium* Art. *K. fysiogr. Sällsk. Lund. Förh.* 7, 119-123.

KYLIN, H. 1938. Beziehung zwischen Generationswechsel und Phylogenie. *Arch. Protistenk.* 90, 432-447.

KYLIN, H. 1944. Die Rhodophyceen der schwedischen Westküste. *Acta Univ. lund.*, N.F., Avd. 2 40 (2), 1-104.

KYLIN, H. 1956. *Die Gattungen der Rhodophyceen*. Gleerup, Lund.

LAMI, R. 1941. L'utilisation des végétaux marins des côtes de France. *Revue int. Bot. appl. Agric. trop.* 21, 653-670.

LAMOUROUX, J. V. F. 1813. Essai sur les genres de la famille de Thalassiophytes, non articulées. *Annls Mus. natn. Hist. nat., Paris*, 20, 115-139; 267-293.

LARSEN, B. & HAUG, A. 1956. Carotene isomers in some red algae. *Acta Chem. scand.* 10, 470-472.

LEE, R. E. 1971a. The pit connections of some lower red algae: ultrastructure and phylogenetic significance. *Br. phycol. J.* 6, 29-38.

LEE, R. E. 1971b. Systemic viral material in the cells of the freshwater red alga *Sirodotia tenuissima* (Holden) Skuja. *J. Cell Sci.* 8, 623-631.

LEMOINE, M. 1910. Répartition et mode de vie du Maerl (*Lithothamnion calcareum*) aux environs de Concarneau (Finistère). *Annls Inst. oceanogr., Monaco*, 1 (3), 1-28.

LEVRING, T. 1953. The marine algae of Australia. I. Rhodophyta: Goniotrichales, Bangiales and Nemalionales. *Ark. Bot.*, A.S. 2, 457-530.

LEVRING, T., HOPPE, H. A. & SCHMIDT, O. J. 1969. *Marine algae: a survey of research and utilization*. Cram. de Gruyter, Hamburg.

LEWIN, R. A. 1962. *Physiology and biochemistry of algae*. Academic Press, New York & London.

LEWIN, R. A. & ROBERTSON, J. A. 1971. Influence of salinity on the form of *Asterocytis* in pure culture. *J. Phycol.* 7, 236-238.

LEWIS, I. F. 1909. The life history of *Griffithsia bornetiana*. *Ann. Bot.* 23, 639-690.

LINNAEUS, C. 1753. *Species plantarum* . . ., 2 vols. Salvius, Stockholm.

LINNAEUS, C. 1758. *Systema naturae* . . ., 2 vols. Salvius, Stockholm.

LINSKENS, H. F. 1963a. Beitrag zur Frage der Beziehung zwischen Epiphyt und Basiphyt bei marinen Algen. *Publ. Staz. zool. Napoli*, 3, 274-293.

LINSKENS, H. F. 1963b. Oberflächenspannung an marinen Algen. *Proc. K. ned. Akad. Wet.* ser. C, 66, 205-217.

LINSKENS, H. F. 1966. Adhäsion von Fortpflanzungszellen benthontischer Algen. *Planta*, 68, 99-110.

LITTLER, M. M. 1972. Biological aspects of the *Porolithon* ridge in tropical Pacific reefs. *J. Phycol.* **8** (suppl.), 10.

LODGE, S. N. 1948. Additions to algal records for the Manx region. *Rep. mar. biol. Stn Port Erin*, **58-60**, 59-62.

LÓPEZ-BENITO, M. 1963. Estudio de la composición química del *Lithothamnium calcareum* (Aresch.) y su aplicación como corrector de terrenos de cultivo. *Investigación pesq.* **23**, 53-70.

LUCAS, A. H. S. & PERRIN, F. 1947. *The seaweeds of South Australia*, vol. 2. Government Printer, Adelaide.

MACFARLANE, C. I. 1964. The seaweed industry of the Maritime Provinces. *Proc. Int. Seaweed Symp.* **4**, 414-419.

MAGNE, F. 1952. La structure du noyau et le cycle nucléaire chez le *Porphyra linearis* Greville. *C. r. hebd. Séanc. Acad. Sci.*, Paris, **234**, 986-988.

MAGNE, F. 1959. Sur le cycle nucléaire du '*Rhodymenia palmata*' (L.) J. Agardh. *Bull. Soc. phycol. Fr.* **5**, 12-13.

MAGNE, F. 1960. Le *Rhodochaete parvula* Thuret (Bangioidée) et sa reproduction sexuée. *Cah. Biol. mar.* **1**, 407-420.

MAGNE, F. 1964a. Recherches caryologiques chez les Floridées (Rhodophycées). *Cah. Biol. mar.* **5**, 461-671.

MAGNE, F. 1964b. La mitose calliblépharidienne de certaines Rhodophycées. *C. r. hebd. Séanc. Acad. Sci.*, Paris, **259**, 3811-3812.

MAGNE, F. 1967a. Sur l'existence, chez les *Lemanea* (Rhodophycées, Némalionales), d'une type de cycle de développment encore inconnu chez les algues rouges. *C. r. hebd. Séanc. Acad. Sci.*, Paris, sér. D, **264**, 2632-2633.

MAGNE, F. 1967b. Sur le déroulement et le lieu de la méiose chez les Lémanéacées (Rhodophycées, Némalionales). *C. r. hebd. Séanc. Acad. Sci.*, Paris, sér. D, **265**, 670-673.

MALLERY, C. H. & RICHARDSON, N. 1971. Disc gel electrophoresis of biliproteins. *Plant Cell Physiol.* **12**, 997-1001.

MANGENOT, G. 1933. Sur les corps irisants de quelques Rhodophycées. *C. r. Séanc. Soc. Biol.* **112**, 659-663.

MARSHALL, S. M., NEWTON, L. & ORR, A. P. 1949. *A study of certain British seaweeds and their utilization in the preparation of Agar.* H.M. Stationery Office, London.

MARTIN, M. T. 1969. A review of life-histories in the Nemalionales and some allied genera. *Br. phycol. J.* **4**, 145-158.

MCBRIDE, D. L. & COLE, K. 1969. Ultrastructural characteristics of the vegetative cell of *Smithora naiadum* (Rhodophyta). *Phycologia*, **8**, 177-186.

MCDONALD, K. 1972. The ultrastructure of mitosis in the marine red alga *Membranoptera platyphylla*. *J. Phycol.* **8**, 156-166.

McLACHLAN, J. 1967. Tetrasporangia in *Asparagopsis armata*. *Br. phycol. Bull.* **3**, 251-252.

MEEUSE, B. J. D. 1962. Storage products. In: *Physiology and biochemistry of algae*, ed. by Lewin, R. A., 289-313. Academic Press, New York & London.

MELCHIOR, H. 1954. Rhodophyta. In: *Engler's Syllabus der Pflanzenfamilien*, *Ed.* 12, ed. by Melchior, H. & Werdermann, E., 123-138. Borntraeger, Berlin.

MEROLA, A. 1956. Le galle nelle alghe Parte I: Storia della cecidogenesi nelle alghe. *Annali Bot.* **25**, 260-281.

MIKAMI, H. 1965. A systematic study of the Phyllophoraceae and Gigartinaeceae from Japan and its vicinity. *Scient. Pap. Inst. algol. Res. Hokkaido Univ.* **5**, 181-285.

MIRANDA, F. 1932. Sobre la homologia de polisporangios y tetrasporangios de la Florideas diplobiontes. *Boln. R. Soc. esp. Histo. nat.* **32**, 191-194.

MIURA, A. 1961. A new species of *Porphyra* and its *Conchocelis*-phase in nature. *J. Tokyo Univ. Fish.* **47**, 305-311.

MOBERLY, R. 1968. Composition of magnesian calcites of algae and pelecypods by electron microprobe analysis. *Sedimentology*, **11**, 61-82.

MULLAHY, J. H. 1952. The morphology and cytology of *Lemanea australis* Atk. *Bull. Torrey bot. Club*, **79**, 393-406; 471-484.

MULLER-STOLL, W. R. 1965. Regeneration bei niederen Pflanzen (in physiologischer Betrachtung). In: *Handbuch der Pflanzenphysiologie*, vol. 15 (2), ed. by Ruhland, W., 92-155. Springer, Berlin.

MURRAY, S. N., SCOTT, J. L. & DIXON, P. S. 1972. The life history of *Porphyropsis coccinea* var. *dawsonii* in culture. *Br. phycol. J.* **7**, 111-122.

MYERS, A. & PRESTON, R. D. 1959. Fine structure in the red algae. III. A general survey of cell-wall structure in the red algae. *Proc. R. Soc.*, ser. B, **150**, 456-459.

MYERS, A., PRESTON, R. D. & RIPLEY, G. W. 1959. An electron microscope investigation into the structure of the Floridean pit. *Ann. Bot.*, N.S. **23**, 257-260.

NAYLOR, G. L. & RUSSELL-WELLS, B. 1934. On the presence of cellulose and its distribution in the cell-walls of brown and red algae. *Ann. Bot.* **48**, 635-641.

NEUSHUL, M. 1972. Functional interpretations of benthic marine algal morphology. In: *Contributions to the systematics of benthic marine algae of the North Pacific*, ed. by Abbott, I. A. & Kurogi, M., 47-73. Japanese Society of Phycology, Kobe.

NEWROTH, P. R. 1972. Studies on life histories in the Phyllophoraceae. I. *Phyllophora truncata* (Rhodophyceae, Gigartinales). *Phycologia*, **10**, 345-354.

NEWTON, L. 1931. *Handbook of the British seaweeds.* British Museum (Natural History), London.

NEWTON, L. 1951. *Seaweed utilization.* Sampson Low, London.

NEWTON, L. M. 1953. Marine algae. *Scient. Rep. John Murray Exped.* 9, 395-420.

NICHOLS, H. W. 1964a. Culture and developmental morphology of *Compsopogon coeruleus. Am. J. Bot.* 51, 180-188.

NICHOLS, H. W. 1964b. Developmental morphology and cytology of *Boldia erythrosiphon. Am. J. Bot.* 51, 653-659.

NICHOLS, H. W. 1965. Culture and development of *Hildenbrandia rivularis* from Denmark and North America. *Am. J. Bot.* 52, 10-15.

NICHOLS, H. W. & LISSANT, E. K. 1967. Developmental studies of *Erythrocladia* Rosenvinge in culture. *J. Phycol.* 3, 6-18.

NICHOLS, H. W., RIDGEWAY, J. E. & BOLD, H. C. 1966. A preliminary ultrastructural study of the freshwater red alga *Compsopogon. Ann. Mo. bot. Gdn,* 53, 17-27.

NORRIS, R. E. 1971. Development of the foliose thallus of *Weeksia fryeana. Phycologia,* 10, 205-213.

NORTH, W. L., STEPHENS, G. C. & NORTH, B. 1972. Marine algae and their relations to pollution problems. *Report of the FAO Technical Conference on marine pollution and its effects on living resources and fishing,* 117-125. FAO, Rome.

O'COLLA, P. S. 1962. Mucilages. In: *Physiology and biochemistry of algae,* ed. by R. A. Lewin, 337-356. Academic Press, New York & London.

OGATA, E. 1954. On the germination of spores in one species of *Hildenbrandtia* in Japan. *Proc. int. bot. Congr.* 8 (17), 106-107.

O'hEOCHA, C. 1962. Phycobilins. In: *Physiology and biochemistry of algae,* ed. by R. A. Lewin, 421-435. Academic Press, New York & London.

O'hEOCHA, C. 1966. Biliproteins. In: *Biochemistry of chloroplasts,* vol. 1, ed. by Goodwin, T. W., 407-422. Academic Press, New York & London.

OKAMURA, K. 1934. *Icones of Japanese algae,* vol. 6. Okamura, Japan.

OLTMANNS, F. 1892. Über die Cultur- und Lebensbedingungen der Meeresalgen. *Jb. wiss. Bot.* 23, 349-440.

OLTMANNS, F. 1898. Zur Entwicklungsgeschichte der Florideen. *Bot. Ztg,* 56, 99-140.

OLTMANNS, F. 1904-1905. *Morphologie und Biologie der Algen,* 2 vols. Fischer, Jena.

PAASCHE, E. & THRONDSEN, J. 1970. *Rhodella maculata* Evans (Rhodophyceae, Porphyridiales) isolated from the plankton of the Oslo Fjord. *Nytt Mag. Bot.* 17, 209-212.

PAPENFUSS, G. F. 1951. Problems in the classification of the marine algae. *Svensk bot. Tidskr.* **45**, 4-11.

PAPENFUSS, G. F. 1955. Classification of the algae. In: *A century of progress in the natural sciences*, 1853-1953, 115-224. California Academy of Sciences, San Francisco.

PAPENFUSS, G. F. 1966. A review of the present system of classification of the Florideophycidae. *Phycologia*, **5**, 247-255.

PERCIVAL, E. & McDOWELL, R. H. 1967. *Chemistry and enzymology of marine algal polysaccharides*. Academic Press, London & New York.

PEYRIÈRE, M. 1969. Infrastructure cytoplasmique du tétrasporocyste de Griffithsia flosculosa (Rhodophycées, Céramiacées). *C. r. hebd. Séanc. Acad. Sci., Paris*, sér. D, **269**, 2332-2334.

PEYRIÈRE, M. 1970. Evolution de l'appareil de Golgi au cours de la tétrasporogenèse de *Griffithsia flosculosa* (Rhodophycées). *C. r. hebd. Séanc. Acad. Sci., Paris*, sér. D, **270**, 2071-2074.

PHILIPPI, R. A. 1837. Beweis, dass die Nulliporen Pflanzen sind. *Arch. Naturgesch.* **3** (1), 387-393.

PHILLIPS, R. W. 1897. On the development of the cystocarp in Rhodymeniales. *Ann. Bot.* **11**, 347-368.

PHILLIPS, R. W. 1925a. On the genera *Phyllophora, Gymnogongrus*, and *Ahnfeldtia* and their parasites. *New Phytol.* **24**, 241-255.

PHILLIPS, R. W. 1925b. On vascular pseudopodia in a species of *Callithamnion. Revue algol.* **2**, 14-18.

PIA, J. 1927. Thallophyta. In: *Handbuch der Palaeobotanik*, vol. 1, ed. by Hirmer, H., 31-136. Oldenbourg, Munich & Berlin.

PIA, J. 1937. Die wichtigsten Kalkalgen des Jungpaläozoikums und ihre geologische Bedeutung. *C. r. Congr. Av. Etude Strat. carb.* **2**, 756-856.

POMEROY, L. R. 1961. Isotopic and other techniques for measuring primary productivity. In: *Proceedings of the Conference on primary productivity measurement*, ed. by Doty, M. S., 97-120. U.S.A.E.C., Honolulu.

POWELL, J. 1964. The life-history of a red alga, *Constantinea*. Ph.D. thesis, University of Washington, Seattle.

PRINGLE, J. D. & AUSTIN, A. P. 1970. The mitotic index in selected red algae *in situ*. II. A supralittoral species, *Porphyra lanceolata* (Setchell & Hus) G. M. Smith. *J. exp. mar. Biol. Ecol.* **5**, 113-137.

PRINTZ, H. 1926. Die Algenvegetation der Trondhjemsfjorden. *K. norske Vidensk. Selsk. Skr.* 1926 (5), 1-274.

PRIOU, M. L. 1962. Recherches sur la structure et la composition des membranes de quelques Rhodophycées. *Annls Sci. nat. (Bot.)*, sér. 12, **3**, 321-406.

PUJALS, C. 1961. Algunas observaciones sobre *Asterocytis ramosa* (C. Ag.) Hamel Rodoficea nueva para Argentina. *Darwiniana*, **12**, 365-377,

RABENHORST, L.   1868.   *Flora europeaea algarum* . . . sect. 3.   Kummerun, Leipzig.

RABINOWITCH, E.   1945-1956.   *Photosynthesis and related processes*, vols. 1, 2 (1), 2 (2).   Interscience, New York.

RAMUS, J.   1969a.   Pit connection formation in the red alga *Pseudogloiophloea*. *J. Phycol.* **5**, 57-63.

RAMUS, J.   1969b.   The developmental sequence of the marine red alga *Pseudogloiophloea* in culture. *Univ. Calif. Publs Bot.* **52**, 1-28.

RAMUS, J.   1971.   Properties of septal plugs from the red alga *Griffithsia pacifica*. *Phycologia*, **10**, 99-103.

RAUNKIAER, C. C.   1934.   *The life forms of plants* . . . . University Press, Oxford.

RAWLENCE, D. J. & TAYLOR, A. R. A.   1970.   The rhizoids of *Polysiphonia lanosa*. *Can. J. Bot.* **48**, 607-611.

REINKE, J.   1878.   Über die Geschlechtspflanzen von *Bangia fuscopurpurea* Lyngb. *Jb. wiss. Bot.* **11**, 274-282.

RENTSCHLER, H. G.   1967.   Photoperiodische Induktion der Monosporenbildung bei *Porphyra tenera* Kjellm. *Planta*, **76**, 65-74.

RICHARDSON, W. N.   1969.   The biology of *Bangia fuscopurpurea* (Dillwyn) Lyngbye. Ph.D. thesis, University of California.

RICHARDSON, W. N.   1970.   Studies on the photobiology of *Bangia fuscopurpurea*. *J. Phycol.* **6**, 216-219.

RICHARDSON, W. N.   1972.   Spore classification in the genera *Bangia* and *Porphyra*. *Br. phycol. J.* **7**, 49-51.

RICHARDSON, W. N. & DIXON, P. S.   1968.   Life history of *Bangia fuscopurpurea* (Dillw.) Lyngb. *Nature, Lond.* **218**, 496-497.

RICHARDSON, W. N. & DIXON, P. S.   1970.   Culture studies on *Thuretellopsis peggiana* Kylin. *J. Phycol.* **6**, 154-159.

ROELOFSEN, P. A.   1959.   *The plant cell wall*. Borntraeger, Berlin.

ROSANOFF, S.   1866.   Recherches anatomiques sur les Mélobésiées. *Mém. Soc. imp. Sci. nat. math. Cherbourg*, **12**, 5-112.

ROSENBERG, T.   1933.   *Studien über Rhodomelaceen und Dasyaceen.* Ohlsson, Lund.

ROSENVINGE, L. K.   1909-1931.   The marine algae of Denmark. Part I. Rhodophyceae. *K. danske Vidensk. Selsk. Skr.*, 7 Raekke, **7**, 1-630.

ROSENVINGE, L. K.   1911.   Remarks on the hyaline unicellular hairs of the Florideae. *Biol. Arb. til E. Warming*, 203-216.

ROSENVINGE, L. K.   1920.   On the spiral arrangement of the branches in some Callithamnieae. *Biol. Meddr*, **2** (5), 1-70.

ROSENVINGE, L. K.   1927.   On mobility in the reproductive cells of the Rhodophyceae. *Bot. Tidsskr.* **40**, 72-79.

ROSENVINGE, L. K.   1929.   *Phyllophora brodiaei* and *Actinococcus subcutaneus*. *Biol. Meddr*, **8** (4), 1-40.

RUENESS, J. 1968. Paraspores from *Plumaria elegans* (Bonnem.) Schmitz. in culture. *Nytt Mag. Bot.* **15**, 220-224.

SAENGER, P., ROWAN, K. S. & DUCKER, S. C. 1969. The water-soluble pigments of the red alga, *Lenormandia prolifera*. *Phycologia*, **7**, 59-64.

SAUVAGEAU, C. 1920. *Utilisation des algues marines*. Doin, Paris.

SAUVAGEAU, C. 1921. Observations biologiques sur le *Polysiphonia fastigiata* Grev. *Recl. Trav. bot. neerl.* **18**, 213-230.

SAUVAGEAU, C. 1926. Sur quelques algues Floridées renferment du brome à l'état libre. *Bull. Stn biol. Arcachon*, **25**, 5-23.

SCAGEL, R. F., BANDONI, R. J., ROUSE, G. E., SCHOFIELD, W. B., STEIN, J. R. & TAYLOR, T. M. C. 1966. *An evolutionary survey of the plant kingdom*. Wadsworth, Belmont.

SCHECHTER, V. 1934. Electrical control of rhizoid formation in the red alga *Griffithsia bornetiana*. *J. gen. Physiol.* **18**, 1-22.

SCHECHTER, V. 1935. The effect of centrifuging on the polarity of an alga, *Griffithsia bornetiana*. *Biol. Bull. mar. biol. Lab.*, *Woods Hole*, **68**, 172-179.

SCHIEWER, U. 1967a. Auxinvorkommen und Auxinstoffwechsel bei mehrzelligen Östseealgen. I. Zur Vorkommen von Indol-3-Essigsäure. *Planta*, **74**, 313-323.

SCHIEWER, U. 1967b. Auxinvorkommen und Auxinstoffwechsel bei mehrzelligen Östseealgen. II. Zur Entstehung von Indol-3-Essigsäure aus Tryptophan unter Berücksichtigung des Einflusses der marinen Bakterienflora. *Planta*, **75**, 152-160.

SCHIEWER, U., KRIENKE, H. & LIBBERT, E. 1967. Auxinvorkommen und Auxinstoffwechsel bei mehrzelligen Östseealgen. III. Die Umsetzung von Indol-3-acetonitril und von Indol-3-Essigsäure. *Planta*, **76**, 52-64.

SCHIEWER, U. & LIBBERT, E. 1965. Indolacetamid—Ein Intermediat der Indolessigsäure-bildung aus Indolacetonitril bei der Alge *Furcellaria*. *Planta*, **66**, 377-380.

SCHILLER, J. 1913. Über Bau, Entwicklung, Keimung und Bedeutung der Parasporen der Ceramiaceen. *Öst. bot. Z.* **63**, 144-149; 203-210.

SCHILLER, J. 1925. Bangiales. In: *Die Süsswasserflora Deutschlands, Österreichs und der Schweiz*, vol. 11, ed. by Pascher, A., 157-164. Fischer, Jena.

SCHIMPER, W. P. 1869. *Traité de paléontologie végétale . . .*, vol. 1. Baillière, Paris.

SCHMITZ, F. 1883. Untersuchungen über die Befruchtung der Florideen. *Sber. Akad. Wiss.* 1883, 215-258.

SCHMITZ, F. 1889. Systematische Übersicht der bischer bekannten Gattungen der Florideen. *Flora, Jena*, **72**, 435-456.

SCHMITZ, F. 1892. Knöllchenartige Auswüchse an den Sprossen einiger Florideen. *Bot. Ztg*, **50**, 624-630.

SCHMITZ, F. & HAUPTFLEISCH, P. 1896-1897. Rhodophyceae. In: *Die naturlichen Pflanzenfamilien* . . ., vol. 1 (2), ed. by Engler, A. & Prantl, K., 298-544. Engelmann, Leipzig.

SCHOTTER, G. 1964. Étude des organes mâles de *Rissoella verruculosa* (Bertol.) J. Ag. (Floridées). *Bull. Inst. oceanogr. Monaco,* 63 (131), 1-31.

SCHOTTER, G. 1968. Recherches sur les Phyllophoracées. *Bull. Inst. océanogr. Monaco,* 67 (1383), 1-99.

SCHUSSNIG, B. 1960. *Handbuch der Protophytenkunde,* vol. 2. Fischer, Jena.

SCHWENKE, H. 1971. Water movement—Plants. In: *Marine ecology,* vol. 1 (2), ed. by Kanne, O., 1091-1092. Wiley-Interscience, London.

SCHWEIGGER, A. F. 1819. *Beobachtungen auf naturhistorischen Reisen, anatomisch-physiologische Untersuchungen über Corallen.* Reimer, Berlin.

SCOTT, J. L. & DIXON, P. S. 1971. The life history of *Pikea californica* Harv. *J. Phycol.* 7, 295-300.

SCOTT, J. L. & DIXON, P. S. 1972a. Ultrastructure of tetrasporogenesis in the marine red alga *Ptilota hypnoides* Harv. *J. Phycol.* 8 (in press).

SCOTT, J. L. & DIXON, P. S. 1972b. Ultrastructure of spermatium liberation in the marine red alga *Ptilota densa* Harv. *J. Phycol.* 8 (in press).

SEGAWA, S. 1936. On the marine algae of Susaki, II. *Scient. Pap. Inst. algol. Res. Hokkaido Univ.* 1, 175-197.

SETCHELL, W. A. 1912. Algae novae et minus cognitae, I. *Univ. Calif. Publs Bot.* 4, 229-268.

SIMON-BICHARD-BREAUD, J. 1971. Un appareil cinétique dans les gametocystes mâles d'un Rhodophycée: *Bonnemaisonia hamifera* Hariot. *C. r. hebd. Séanc. Acad. Sci., Paris,* sér. D, 273, 1272-1275.

SIMONS, R. H. 1960. Notes on *Aristothamnion purpuriferum* (Kütz.) J. Ag. *Bothalia,* 7, 201-205.

SIRODOT, S. 1884. *Les Batrachospermes Organisation, fonctions, développement, classification.* Masson, Paris.

SJÖSTEDT, L. G. 1926. Floridean studies. *Acta Univ. lund.,* N.F., Avd. 2, 22 (4), 1-95.

SKUJA, H. 1933. Untersuchungen über die Rhodophyceen des Süsswassers. III. *Batrachospermum breutelii* Rbh und seine Brutkörper. *Arch. Protistenk.* 80, 357-360.

SKUJA, H. 1938. Die Süsswasserrhodophyceen der Deutschen Limnologischen Sunda-Expedition. *Arch. Hydrobiol. Suppl.* 15, 603-637.

SKUJA, H. 1939. Versuch einer systematischen Einteilung der Bangioideen oder Protoflorideen. *Acta Horti bot. Univ. latv.* 11/12, 23-40.

SMITH, G. M. 1938. *Cryptogamic Botany*, vol. 1. McGraw-Hill, New York & London.

SMITH, G. M. 1944. *Marine algae of the Monterey Peninsula.* Stanford University Press, Stanford.

SMITH, G. M. 1955. *Cryptogamic botany*, vol. 1. McGraw-Hill, New York & London.

SMITH, G. M. & HOLLENBERG, G. J. 1943. On some Rhodophyceae from the Monterey Peninsula, California. *Am. J. Bot.* 30, 211-222.

SOLMS-LAUBACH, H. 1881. Die Corallinenalgen des Golfes von Neapel. *Fauna Flora Golfes Neapel*, 4, 1-64.

SOMMERFELD, M. R. & NICHOLS, H. W. 1970a. Comparative studies in the genus *Porphyridium* Naeg. *J. Phycol.* 6, 67-78.

SOMMERFELD, M. R. & NICHOLS, H. W. 1970b. Developmental and cytological studies of *Bangia fuscopurpurea* in culture. *Am. J. Bot.* 57, 640-648.

SOUTH, G. R., HOOPER, R. G. & IRVINE, L. M. 1972. The life history of *Turnerella pennyi* (Harv.) Schmitz. *Br. phycol. J.* 7, 221-233.

STACKHOUSE, J. 1795-1801. *Nereis Britannica.* . . . Hazard, Bath.

STACKHOUSE, J. 1809. Tentamen marino-cryptogamicum. *Mém. Soc. imp. Nat. Moscou*, 2, 50-97.

STACKHOUSE, J. 1816. *Nereis Britannica* . . ., 2nd edition. Collingwood, Oxford.

STARMACH, K. 1952. The reproduction of the freshwater Rhodophyceae *Hildenbrandia rivluaris* (Liebm.) J. Ag. *Acta Soc. bot. Polon.* 21, 447-474.

STEENSTRUP, J. J. S. 1842. *Über den Generationswechsel.* . . . Luno, Copenhagen.

STEINECKE, F. 1925. Der Stammbaum der Algen nach serio-diagnostischen Untersuchungen dargestellt. *Bot. Archiv.* 10, 82-208.

STERNBERG, G. K. 1833. *Flora der Vorwelt* . . ., vol. 2 (5-6).

STEWART, J. G. 1968. Morphological variation in *Pterocladia pyramidale. J. Phycol.* 4, 76-84.

STOLOFF, L. 1962. Algal classification—an aid to improved industrial utilization. *Econ. Bot.* 16, 86-94.

STOLOFF, L. & SILVA, P. 1957. An attempt to determine possible taxonomic significance of the properties of water-extractable polysaccharides in red algae. *Econ. Bot.* 11, 327-330.

VON STOSCH, H. A. 1965. The sporophyte of *Liagora farinosa* Lamour. *Br. phycol. Bull.* 2, 486-496.

STRASBURGER, E. 1882. *Über den Bau und das Wachstum der Zellhäute.* Fischer, Jena.

SUNDENE, O. 1962. Reproduction and morphology in strains of *Antithamnion boreale* orginating from Spitzbergen and Scandinavia. *Skr. norske Vidensk-Akad., Mat.-naturv. Kl.* 1962 (5), 1-19.

SUNESON, S.   1937.   Studien über die Entwicklungsgeschichte der Corallinaceen.   *Acta Univ. lund.*, N.F., Avd. 2, **33** (2), 1-102.

SUNESON, S.   1950.   The cytology of bispore formation in two species of *Lithophyllum* and the significance of the bispores in the Corallinaceae. *Bot. Notiser,* 1950, 429-450.

SVEDELIUS, N.   1908.   Über den Bau und die Entwicklung der Florideengattung *Martensia*. *K. Svenska Vetensk-Akad. Handl.* **43** (7), 1-101.

SVEDELIUS, N.   1911.   Über den Generationswechsel bei *Delesseria sanguinea.* Svensk bot. *Tidskr.* **5**, 260-324.

SVEDELIUS, N.   1914.   Über die Tetradenteilung in den viel kernigen Tetrasporangiumanlagen bei *Nitophyllum punctatum.* *Ber. dt. bot. Ges.* **32**, 106-116.

SVEDELIUS, N.   1915.   Zytologisch-entwicklungsgeschichtliche Studien über *Scinaia furcellata.* *Nova Acta R. Soc. Scient. upsal.,* ser. 4, **4** (4), 1-55.

SVEDELIUS, N.   1927.   Alternation of generations in relation to reduction division. *Bot. Gaz.* **83**, 362-384.

SVEDELIUS, N.   1931.   Nuclear phases and alternation in the Rhodophyceae. *Beih. bot. Zbl.* **48**, 38-59.

SVEDELIUS, N.   1933.   On the development of *Asparagopsis armata* Harv. and *Bonnemaisonia asparagoides* (Woodw.) Ag. *Nova Acta R. Soc. Scient. upsal.,* ser. 4, **9** (1), 1-61.

SVEDELIUS, N.   1937.   The apomeiotic tetrad division in *Lomentaria rosea.* *Symb. bot. upsal.* **2** (2), 1-54.

SVEDELIUS, N.   1939.   Anatomisch-entwicklungsgeschichtliche Studien über die Florideengattung *Dermonema* (Grev.) Harv. *Bot. Notiser,* 1939, 21-39.

TAYLOR, W. R.   1937.   *Marine algae of the northeastern coast of North America.* University of Michigan Press, Ann Arbor.

TERBORGH, J. & THIMANN, K.   1964.   Interactions between daylength and light intensity in the growth and chlorophyll content of *Acetabularia crenulata. Planta,* **63**, 83-98.

THAXTER, R.   1900.   Note on the structure and reproduction of *Compsopogon. Bot. Gaz.* **29**, 259-267.

THURET, G.   1855.   Recherches sur la fécondation des Fucacées et les anthéridies des algues, II. *Annls Sci. nat. (Bot.),* sér. 4, **3**, 5-28.

THURET, G. & BORNET, E.   1878.   *Études phycologiques.* Masson, Paris.

TOBLER, F.   1904.   Über Eigenwachstum der Zelle und Pflanzenform. *Jb. wiss. Bot.* **39**, 527-580.

TOBLER, F.   1906.   Über Regeneration und Polarität sowie verwandte Wachstumsvorgange bie *Polysiphonia* und anderen Algen. *Jb. wiss. Bot.* **42**, 461-502.

TOKIDA, J.   1958.   A review on galls in seaweeds. *Bull. Jap. Soc. Phycol.* **6**, 93-99.

TOKIDA, J. & YAMAMOTO, H. 1965. Syntagmatic germination of tetraspores in *Pachymeniopsis yendoi*. *Phycologia*, **5**, 15-20.

TSENG, C. K. 1946. Phycocolloids: useful seaweed polysaccharides. In: *Colloid Chemistry*, vol. 6, ed. by Alexander, J. New York.

TSENG, C. K. & CHANG, T. J. 1954. Studies on *Porphyra*. I. Life history of *Porphyra tenera* Kjellm. *Acta bot. sin.* **3**, 287-302.

TSENG, C. K. & CHANG, T. J. 1955. Studies on *Porphyra*. III. Sexual reproduction of *Porphyra*. *Acta bot. sin.* **4**, 153-166.

TURNER, D. 1802. *A synopsis of the British fuci*, 2 vols. Bush, Yarmouth.

UMEZAKI, I. 1969. The germination of tetraspores of *Hildenbrandia prototypus* Nardo and its life history. *J. Jap. Bot.* **44**, 17-28.

VINOGRADOV, A. P. 1953. *The elementary chemical composition of marine organisms*. New Haven.

WAALAND, S. D. & CLELAND, R. 1972. Development in the red alga, *Griffithsia pacifica*: control by internal and external factors. *Planta*, **105**, 196-204.

WAALAND, S. D., DUFFIELD, E. S. & CLELAND, R. E. 1971. *Griffithsia pacifica*: a new developmental system. *J. Phycol.* **7** (suppl.), 7.

WAALAND, S. D., WAALAND, J. R. & CLELAND, R. 1972. A new pattern of plant cell elongation: bipolar band growth. *J. Cell Biol.* **54**, 184-190.

WAERN, M. 1952. Rocky-shore algae in the Oregrund Archipelago. *Acta phytogeogr. suec.* **30**, i-xvi, 1-298.

WARDLAW, C. W. 1955. *Embryogenesis in plants*. London.

WEST, J. A. 1968. Morphology and reproduction of the red alga *Acrochaetium pectinatum* in culture. *J. Phycol.* **4**, 89-99.

WEST, J. A. 1969. The life histories of *Rhodochorton purpureum* and *R. tenue* in culture. *J. Phycol.* **5**, 12-21.

WEST, J. A. 1970. The life history of *Rhodochorton concrescens* in culture. *Br. phycol. J.* **5**, 179-186.

WEST, J. A. 1972. The life history of *Petrocelis franciscana*. *Br. phycol. J.* **7**, 111-222.

WEST, J. A. & NORRIS, R. E. 1966. Unusual phenomena in the life histories of Florideae in culture. *J. Phycol.* **2**, 54-57.

WESTBROOK, M. A. 1928. Contributions to the cytology of tetrasporic plants of *Rhodymenia palmata* (L.) Grev. and some other Florideae. *Ann. Bot.* **42**, 149-179.

WESTBROOK, M. A. 1930. *Compsothamnion thuyoides* (Smith) Schmitz. *J. Bot., Lond.* **68**, 353-364.

WESTBROOK, M. A. 1935. Observations on nuclear structure in the Florideae. *Beih. bot. Zbl.*, **A53**, 654-585.

WHITE, E. B. & BONEY, A. D. 1969. Experiments with some endophytic and endozoic *Acrochaetium* species. *J. expl. mar. Biol. Ecol.* **3**, 246-274.

WILCE, R. T. 1967. Heterotrophy in arctic sublittoral seaweeds: an hypothesis. *Botanica mar.* **10**, 185-197.

WILKE, H. 1929. Die Phylogenie der Rhodophyceae. *Bot. Arch.* **26**, 1-85.

WILLE, N. 1900. Algologische Notizen, III. *Nyt Mag. Naturv.* **38**, 7-10.

WOELKERLING, W. J. 1971. Morphology and taxonomy of the *Audouinella* complex (Rhodophyta) in Southern Australia. *Aust. J. Bot.* Suppl. **1**, 1-91.

WOLK, C. P. 1968. Role of bromine in the formation of the refractile inclusions of the vesicle cells of the Bonnemaisoniaceae (Rhodophyta). *Planta*, **78**, 371-378.

YABU, H. 1967. Nuclear division in *Bangia fuscopurpurea* (Dillw.) Lyngbye. *Bull. Fac. Fish. Hokkaido Univ.* **17**, 163-164.

YABU, H. 1969. Observation on chromosomes in some species of *Porphyra*. *Bull. Fac. Fish. Hokkaido Univ.* **19**, 239-243.

YABU, H. 1970. Cytology in two species of *Porphyra* from the stipes of *Nereocystis luetkeana* (Mert.) Post. et Rupr. *Bull. Fac. Fish. Hokkaido Univ.* **20**, 243-251.

YABU, H. 1971. Observation on chromosomes in some species of *Porphyra*, II. *Bull. Fac. Fish. Hokkaido Univ.* **21**, 253-258.

YABU, H. 1972. Observation on chromosomes in some species of *Porphyra*, III. *Bull. Fac. Hokkaido Univ.* **22**, 261-266.

YABU, H. & TOKIDA, J. 1963. Mitosis in *Porphyra*. *Bull. Fac. Fish. Hokkaido Univ.* **14**, 131-136.

YAMANOUCHI, S. 1906. The life-history of *Polysiphonia violacea*. *Bot. Gaz.* **41**, 425-433; **42**, 401-449.

YAPHE, W. 1959. The determination of K-carrageenin as a factor in the classification of the Rhodophyceae. *Can. J. Bot.* **37**, 751-757.

INDEX